Prais[illegible]

"What C-level executives read to keep [illegible] decisions. Timeless classics for indispensable knowledge." - Richard Costello, Manager of Corporate Marketing Communication, General Electric

"Want to know what the real leaders are thinking about now? It's in here." - Carl Ledbetter, SVP and CTO, Novell Inc.

"Priceless wisdom from experts at applying technology in support of business objectives." - Frank Campagnoni, CTO, GE Global Exchange Services

"Unique insights into the way the experts think and the lessons they've learned from experience." - MT Rainey, Co-CEO, Young & Rubicam/Rainey Kelly Campbell Roalfe

"A must-read for anyone in the industry." - Dr. Chuck Lucier, Chief Growth Officer, Booz-Allen & Hamilton

"Unlike any other business books, *Inside the Minds* captures the essence, the deep-down thinking processes, of people who make things happen." - Martin Cooper, CEO, Arraycomm

"A must-read for those who manage at the intersection of business and technology." - Frank Roney, General Manager, IBM

"A great way to see across the changing marketing landscape at a time of significant innovation." - David Kenny, Chairman and CEO, Digitas

"An incredible resource of information to help you develop outside the box..." - Rich Jernstedt, CEO, Golin/Harris International

"A snapshot of everything you need to know..." - Larry Weber, Founder, Weber Shandwick

"Great information for both novices and experts." - Patrick Ennis, Partner, ARCH Venture Partners

"The only useful way to get so many good minds speaking on a complex topic." - Scott Bradner, Senior Technical Consultant, Harvard University

"Must-have information for business executives." - Alex Wilmerding, Principal, Boston Capital Ventures

www.Aspatore.com

Aspatore Books is the largest and most exclusive publisher of C-Level executives (CEO, CFO, CTO, CMO, Partner) from the world's most respected companies and law firms. Aspatore annually publishes a select group of C-Level executives from the Global 1,000, top 250 law firms (Partners & Chairs), and other leading companies of all sizes. C-Level Business Intelligence™, as conceptualized and developed by Aspatore Books, provides professionals of all levels with proven business intelligence from industry insiders – direct and unfiltered insight from those who know it best – as opposed to third-party accounts offered by unknown authors and analysts. Aspatore Books is committed to publishing an innovative line of business and legal books, those which lay forth principles and offer insights that when employed, can have a direct financial impact on the reader's business objectives, whatever they may be. In essence, Aspatore publishes critical tools – need-to-read as opposed to nice-to-read books – for all business professionals.

Inside the Minds

The critically acclaimed *Inside the Minds* series provides readers of all levels with proven business intelligence from C-level executives (CEO, CFO, CTO, CMO, partner) from the world's most respected companies. Each chapter is comparable to a white paper or essay and is a future-oriented look at where an industry/profession/topic is heading and the most important issues for future success. Each author has been carefully chosen through an exhaustive selection process by the *Inside the Minds* editorial board to write a chapter for this book. *Inside the Minds* was conceived in order to give readers actual insights into the leading minds of business executives worldwide. Because so few books or other publications are actually written by executives in industry, *Inside the Minds* presents an unprecedented look at various industries and professions never before available.

INSIDE THE MINDS

Making Critical Technology Decisions

Leading CTOs & CIOs on Identifying Opportunities, Calculating Return on Investments, and Aligning Technology with Business Goals

BOOK IDEA SUBMISSIONS

If you are a C-level executive or senior lawyer interested in submitting a book idea or manuscript to the Aspatore editorial board, please e-mail authors@aspatore.com. Aspatore is especially looking for highly specific book ideas that would have a direct financial impact on behalf of a reader. Completed books can range from 20 to 2,000 pages—the topic and "need-to-read" aspect of the material are most important, not the length. Include your book idea, biography, and any additional pertinent information.

SPEAKER SUBMISSIONS FOR CONFERENCES

If you are interested in giving a speech for an upcoming ReedLogic conference (a partner of Aspatore Books), please e-mail the ReedLogic Speaker Board at speakers@reedlogic.com. If selected, speeches are given over the phone and recorded (no travel necessary). Due to the busy schedules and travel implications for executives, ReedLogic produces each conference on CD-ROM, then distributes the conference to bookstores and executives who register for the conference. The finished CD-ROM includes the speaker's picture with the audio of the speech playing in the background, similar to a radio address played on television.

INTERACTIVE SOFTWARE SUBMISSIONS

If you have an idea for an interactive business or software legal program, please e-mail software@reedlogic.com. ReedLogic is specifically seeking Excel spreadsheet models and PowerPoint presentations that help business professionals and lawyers accomplish specific tasks. If idea or program is accepted, product is distributed to bookstores nationwide.

Published by Aspatore Inc.

For corrections, company/title updates, comments, or any other inquiries, please e-mail store@aspatore.com.

First Printing, 2005
10 9 8 7 6 5 4 3 2 1

ISBN 1-59622-394-4
Library of Congress Control Number: 2005938714

Inside the Minds Managing Editor, Laura Kearns, Edited by Eddie Fournier, Proofread by Brian Denitzio

Making Critical Technology Decisions

Leading CTOs & CIOs on Identifying Opportunities, Calculating Return on Investments, and Aligning Technology with Business Goals

CONTENTS

Change Management: A Contact Sport

Ralph E. Loura

Vice President and Chief Information Officer

Symbol Technologies Inc.

Establishing a Vision and Goals

Having spent twenty years in the technology sector and fifteen of that in various roles in or surrounding information technology (IT) organizations, I have come to the conclusion that IT organizations exist to fulfill one primary and one secondary goal in any company. The primary goal is to *increase corporate productivity*, and the secondary goal is to *enhance the customer experience.* I have never come across a valid IT mission statement that could not ultimately be boiled down to one of these two elemental goals. Once stated, these seem rather intuitive and perhaps even overly simplified; however, I can attest from personal experience that, far too often, technology initiatives proceed to completion that would never pass the productivity or customer experience filters.

In my role as chief information officer (CIO) of Symbol Technologies, my personal vision is to develop a team of balanced individuals who work in an environment of enlightened empowerment in support of the aforementioned goals. By balanced, I am referring to people who are equally adept at dealing in the realm of people, processes, or technology. Technology-driven organizations can get caught up in the *technology for technology's sake* race, while process-driven organizations often devolve into a morass of red tape and bureaucracy. I have yet to come across a purely people-driven IT function; however, it would surely have its own perils. One of my goals is to develop a team that is skilled in each of the three areas referenced and that strives to keep them in harmony.

IT organizations exist at the discretion of the businesses they support. My role as CIO is to be sufficiently connected to each of my constituent business owners to establish an understanding of their key business problems, what business processes are at play, and who is impacted by those problems. IT leaders are largely facilitators and mediators. Truly skilled IT leaders are better at asking questions than they are at providing on-the-spot answers.

Occasionally, problems (and therefore solutions) are predominantly technology-related; however this is the exception and not the rule. Ultimately, business is about people and processes, with technology trailing behind.

The only areas where today's CIO makes decisions directly are those that involve traditional IT systems and core elements of an organization's enterprise architecture. This means everything from network solutions and server and storage platforms to application servers. Even this area requires collaboration in some environments, and is a particular challenge with being the CIO of a company that makes IT products.

Making a Financial Impact

There are several areas in which IT functions have a direct financial impact and add value to the company. The first is in the core area of business productivity. We have to deliver on our core mission, which is to enhance the productivity of our associate base and support the activities of their end customers. Additionally, IT functions are increasingly being sought out as partners in more revolutionary business change. At Symbol, we embarked on a business transformation initiative roughly two years ago. Our business transformation program was a set of business-driven initiatives focused on business process reengineering across all functions in the firm. IT served as a co-leader and partner to the business in these efforts with a variety of goals ranging from inventory reduction to customer satisfaction improvement. All in all, if an IT function is having trouble defining where it is creating positive financial impact, there is something fundamentally wrong with the function.

One of the largest areas of discretionary spending in many IT shops is with application integration consultants; this is particularly true in cycles of heavy application build-out. Poor focus and control in this area is perhaps the greatest risk to erosion of value created. Strong IT leaders must also be able vendor relations managers and adept at managing outside services elements of their budgets.

Lastly, it is simply good practice to perform due diligence on recurring spend categories like network services, equipment leasing, and so on. For smaller firms, the services of large research firms in this area—or more recently, firms specializing in request for proposal/request for information facilitation—may be well worth the cost in order to ensure that you are as effective as possible in these areas, thus maintaining a competitive overall spend rate.

Being a Good CIO: An Art Form

The modern-era CIO needs to be a Renaissance person. Decades ago, the CIO was either a financial manager who focused on controlling IT spend or a technology leader whose focus was biased toward technology strategy and management. These days, a CIO needs to have a broadly balanced set of attributes and a broader focus grounded in the needs of the business.

The art of being a CIO is the art of finding a balance between people, process, and technology. From one perspective, the CIO is the translator who maps the corporate vision and goals against the reality of the technology landscape. From another perspective, the CIO is the author of the enterprise architecture. From yet another perspective, the CIO is the firm's chief agent of change. All are valid when integrated with the others; none are valid on their own. Ultimately, every technology decision should start and end with a business need.

Increasingly, CIOs are expected to be agents of change. All too often, leaders underestimate the effort and constancy of purpose required to effect real change. The efforts leading up to a decision and then the decision itself can feel uncannily like change itself, when in fact it is simply the precursor. Many people are familiar with the old children's riddle of five frogs on a log. Four frogs decide to jump in the pond. The question: How many frogs are left? The answer: Five, because there's a difference between deciding and doing. Change is a contact sport, and it is an endurance sport. Just because a decision has been made doesn't mean it has been realized. The difference between deciding and doing is significant. Good leaders understand this and work toward real change on a continuous basis.

A CIO should show up to work every day with the mentality that he or she needs to earn the job each day and justify his or her organization to the business it supports. IT is a substantial expense to any business. If a CIO's focus isn't on delivering value and retaining a competitive edge, IT can quickly become irrelevant to a company and the CIO irrelevant to the IT function.

Strategies for Success

In order to be successful when making technology decisions for a company, a CIO needs business acumen, tact, diplomacy, and vendor management skills. Increasingly, we live in a codependent world. The key to success is the ability to perform a multitude of activities concurrently and in real time while exerting both direct and indirect influence across different people and organizations.

I have developed several strategies for success that I use on a regular basis. One is the use of a progressive review process. We know that all of the initial decisions we make won't always be right. As we gain experience and knowledge, we also gain greater insight and we are able to hone in on better decisions. The key is to reduce risk up front. We don't bet the farm early.

At Symbol, we employ a phase-gated review methodology that allows us to do requirements management, blueprinting, and development in phases with formalized review "gates" between each phase. This allows us to be sure there is appropriate review and feedback as we move along a particular path.

Another process I employ is shared risk taking, both internally and externally. A good indicator that a project may not be appropriate is if we're having difficulty finding a business sponsor who is willing to put their budget on the line or commit future returns on the success of the project. When dealing with managed service contractors or other solutions players in the business, it's common to set up similar shared risk-taking structures that include both service level agreements and client satisfaction measures in the compensation formula.

Even the best of intentions and processes can be subverted by someone with a strong technology or vendor bias. One of the challenges technology leaders face is the issue of people who are entrenched in their positions seemingly beyond objective reason. The armchair technologist who has made up his or her mind that a particular solution is right presents a problem. The most effective strategy I've seen is to attempt to co-op the entrenched parties. As opposed to letting people sit on the outside and throw rocks at the solution, I invite them to be part of the architecture team, to be responsible for the outcome, and to be an owner in the process of making tradeoffs and selecting the best solution. This approach tends to broaden the person's viewpoint, because he or she realizes that their solution may not work when viewed in light of integration challenges or scalability issues. This method addresses the issue of that particular individual and often makes that person a proponent who can help you sell the solution once it has been resolved.

I try very hard to keep conversations away from technology and focused on the business. I also try not to allow projects and programs, once they're up and running, to be categorized or viewed as IT programs. If there's an implementation of a system in support of a business process, it needs to be viewed as a business initiative with an IT component, not the other way around.

Calculating ROI

When calculating return on investment (ROI) with respect to technology decisions, we take the view that they're not technology decisions, but rather business decisions. We look at all aspects of a decision. If we're implementing a new approach to supply chain management in operations in an attempt to lower inventory levels or improve forecast accuracy, we begin with the business issue we're trying to solve, add up all the costs, and try to understand the net impact on the people, process, and technology. We build our ROI in that regard.

We evaluate previously purchased technologies on a consistent basis to monitor their worth. It's often difficult to separate other events in the environment that have occurred along the way. We attempt to validate the things we expected to happen, such as whether we stayed on time, on spec, and on budget, and ask if we had the net impact at the end that we expected.

In order for us to consider a technology investment worthwhile, it has to satisfy multiple dimensions of analysis. One of these dimensions is a hard ROI, but we also ask if it makes strategic sense in terms of the overall technology roadmap of the company from an IT perspective and a market perspective.

Challenges and Misconceptions

The most difficult types of technology decisions I am faced with involve multi-year or multi-quarter timelines and large expenses, because the stakes are so high. I prefer something I can roll incrementally, where I can see early feedback and gain ROI faster. The decisions that require big bang conversion after a large amount of activity are the ones with the most risk.

One of the biggest misconceptions about the CIO role is that the CIO makes technology decisions. My role is largely to play mediator, broker, and provider of information. Once decisions are made, my role is to communicate up, down, and across the organization what those decisions are and why they were made. The CIO helps the company arrive at a consensus and communicates a decision to ensure that it is executed.

One of the most challenging aspects of being a CIO is time management, and the only way to address it is to be proactive. First of all, get away from your e-mail for big chunks of time each day. Don't start the day with e-mail, and don't let e-mail drive your activity for the day. Check your activities each day against your higher-level priorities and defer (or drop) items that are not supportive of your personal and organizational agenda.

The CIO is expected to operate at a very strategic level and at the same time have a great deal of detailed understanding. Internally, we talk about the blender versus the helicopter. You want to be hovering above the fray, assessing what's going on and making decisions based on a larger scope and long-term plan, rather than getting caught in the blender running as fast as you can to try and stay above the blades.

The Team

As CIO, when making technology decisions I work closely with the functional leaders, including the senior vice presidents responsible for various aspects of the business, the chief financial officer, the chief technology officer, and the chief executive officer.

A common mistake of technology leaders in past decades was to keep technology problems at the forefront of discussions. Now, we have the attitude that there are no technology decisions and no technology projects, only business decisions and business projects. Business decisions often carry a technology component; the technology issues should be secondary.

When we set goals as a team, we follow a pretty standard formalized annual process. We start with our vision, mission, goals, and initiatives statement. We effectively segment that statement to develop initiatives for our sub-organizations, down to the associate level. We address metrics. We don't have a measurement that focuses on implementing release X by date Y. Instead, we focus on improving the business productivity of X department by Y percent via deployment of Z solution. We keep our metrics and measures focused on business results and not on technology elements.

I tell my team members to set the technology debate aside and focus on what's important to the business. We have to keep the business's needs at the forefront and play the role of third-party advisor, not act as an advocate for any particular technology or cause. When you advocate, it puts people in a defensive position, and they begin to argue and resist. We realize that everyone's goals are ultimately the same and that enlightened users will operate in the area of self-interest and support the decision that makes the most sense for the company.

The best piece of advice I have ever received came from a former manager of mine. The advice was simply that change management is a contact sport, and just because something makes sense to the team that spent nine months developing a solution, it may not make sense to the individual who just found it on their desk that morning. Change is a contact sport, and it requires interacting with people, making them comfortable, and helping them see the rationale behind various decisions.

Changes in the Role of the CIO

In the past five to ten years, the CIO role has been elevated to some degree by an understanding that the technology and application environment within a company supports business process reengineering and business process transformation. CIOs have been brought out of the wiring closet and the data center and into a role in the boardroom. This change has required much adaptation and evolution in the organization and the person in the role itself.

Ralph E. Loura is vice president and chief information officer of Symbol Technologies, where he is responsible for the company's worldwide use of information technology. In this role, Mr. Loura leads Symbol's information technology team and directs the company's strategic technology initiatives to build a systems infrastructure that can drive new efficiencies and productivity into the company's operations.

Since joining Symbol in 2003, Mr. Loura has led the company's implementation of vital systems infrastructure and process improvements as part of its business transformation program, enabling Symbol to transform its core operations into a strategic competitive advantage for the company.

Prior to joining Symbol, Mr. Loura held a variety of information technology leadership positions at Cisco Systems, Lucent Technologies Research, and AT&T Network Systems. Most recently, at Cisco Systems, Mr. Loura was responsible for overall information technology infrastructure for a large client base in geographically distributed locations throughout North and South America, as well as Europe. Mr. Loura received a master's degree in computer science from Northwestern University and a bachelor's degree in computer science and mathematics from Saint Joseph's College.

Managing Technology's Impact on People and Processes

Thomas H. Murphy

Senior Vice President and Chief Information Officer

AmerisourceBergen

What is a CIO?

The role of the chief information officer (CIO) has evolved over the years from a technical resource to a strategic resource who has a macro view of the company. The effective CIO sees across and influences all business units and departmental silos, and thus shares a view of the company that generally only the chief executive officer has enjoyed in the past. The CIO must have business acumen, be able to build relationships, and be a practical visionary. He or she should be a member of the executive leadership team instead of simply being a technical advisor to the team. The effective CIO must balance his or her time between the operational and strategic aspects of the company. How much is spent on either depends on the quality of the information technology (IT) staff and where the business is in their lifecycle. The CIO should be able to improve profitability by understanding the strategies and drivers that are moving the company as well as by understanding the problems or opportunities the company has and applying technology to those problems or opportunities. He or she needs to be able to show the value new technology is going to bring to the company.

The CIO should work to build a high-performance team that deeply engages with the business to enable competitive advantage, cost improvement, and cost efficiency. The team must be willing to take risks. They must possess the business alignment to bring powerful and applicable technology into being. The CIO should hire people smarter than the CIO and then give them the freedom to execute. This allows the CIO to spend the appropriate time on strategic opportunities.

Challenges of a CIO

The most challenging aspect of the CIO position is getting the executive management team to work well together and to align the IT investment with the business priorities. Everybody wants to do a great job, but very few people look at the strategic horizon. They see what they need within their department, and it's difficult to get people to look across the enterprise and compromise to ensure value to the entire company. In order to maximize the value IT can drive, the company must manage that IT investment at an enterprise level. The CIO must pull everyone together and help the organization to understand that not every idea can be prioritized and executed. Once initiatives are underway, it is difficult to manage the change and to convince people that change is best for the company in the long run. It is difficult for all of us to adjust to change, even when shown that change will make the company more successful.

Working as a Team

When making technology decisions with other executives at a company, the CIO should put together a business opportunity council (BOC) or steering committee. The top executives in the company must be deeply engaged in technical decisions, because the right application of technology is really a business decision, not a technical one. In addition, it is critical for the leadership of the company to recognize that the IT asset is not free and not unlimited. Priorities must be established and compromises made by the executive team, not the CIO. Those compromises must be communicated to the organization. In order to make those compromises, the BOC should have request criteria. Here is a suggested criteria list:

- Project must be linked to a corporate program or strategy, be a regulatory mandate, or add significant value to the business by demonstrating one or more of the following:
    a. Increased profitability
    b. Increased revenue
    c. Decreased expense
    d. Risk mitigation
    e. Operational efficiency
    f. Customer satisfaction
- Must indicate total cost (resources, hardware, software, ongoing support costs).
- Short-term solutions must specify percent throwaway once the long-term solution is implemented.
- Must be sponsored by a BOC member.

The BOC's charter should include:

- Initiate corporate IT programs to enable the business strategic direction.
- Review and approve changes to the established program roadmaps.
- Monitor programs: priority, scope changes, and return on investment.
- Ensure that a corporate view is considered for all major programs.
- Implement an enterprise-driven prioritization process, thus managing departmental versus corporate priority conflicts.
- Filter unplanned requests for their strategic fit and bottom line contribution.

In addition to the CIO, the council should include the president, the chief financial officer, the chief operating offer, the chief human resource officer, and the chief procurement officer. Those executives should look at the technology from a business perspective and apply a series of filters to determine whether the technology fits into the values and strategies of the company. The council must also determine if the technology will work within the limited human and financial resources of the company. It is crucial that each member of the council understand each other's priorities, because compromises will be required. For instance, there may be a case where a new customer application is important to the sales executive but there is competition for IT resources to complete an operational system that supports the warehouse. The executive team must make a choice between the two. It is equally important that the lines of communication are open between council members.

Financial Impact

There are many ways a CIO can have a direct financial impact on a company. The most obvious include the practical and effective application of technology and the cost-effective management of the IT environment. A more nuanced way is to build an organization of process experts who can work across departmental silos to smooth transaction and customer flows, making the company more efficient and "easier to do business with." Often, this process analysis and restructure has no technical requirements and thus can be achieved at relatively low cost.

When applying new technologies, a large part of success is finding a way to show the company what's possible. It is very hard to articulate to executives a world in which they have not lived before. The CIO must have a vision and help executives imagine the possibilities of that vision. In order to align effectively and thus drive the most financial impact, it is useful to create an alignment "radar" that the business and IT can agree represents the core business strategies and IT's response.

The IT Strategic Radar Screen

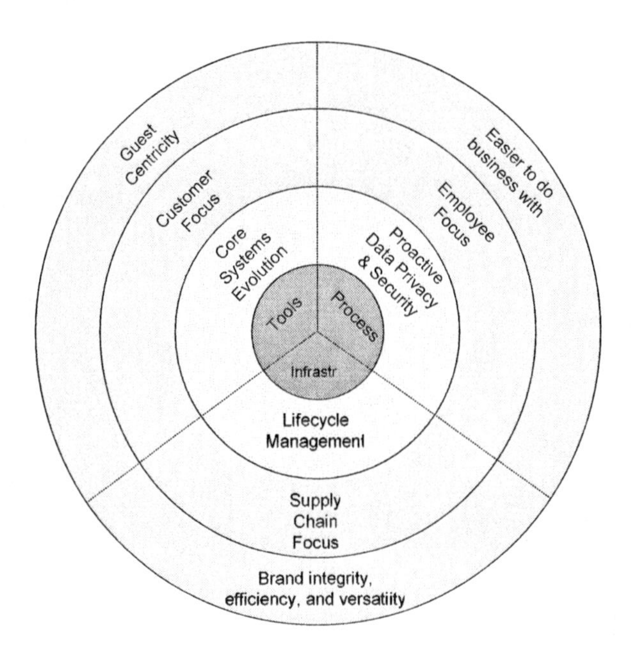

Successful Strategies

There are several specific strategies or methodologies a CIO can use that lead to success on a regular basis. The first is to hire smart people. The CIO who doesn't want to surround him or herself with employees who are smarter than the CIO will ultimately fail or will be relegated to a tactical role. It is necessary to hire incredibly smart people when running a large organization. The newly hired CIO coming into an organization must do a rapid assessment of the quality of the talent against the expectations of the company, and then make changes very quickly. He or she should spend a lot of time on the human resource side of leadership.

Another strategy is to implement roadmapping. Here is an example of the "as-is" state and the "to-be" state diagramming that sets the stage for roadmapping:

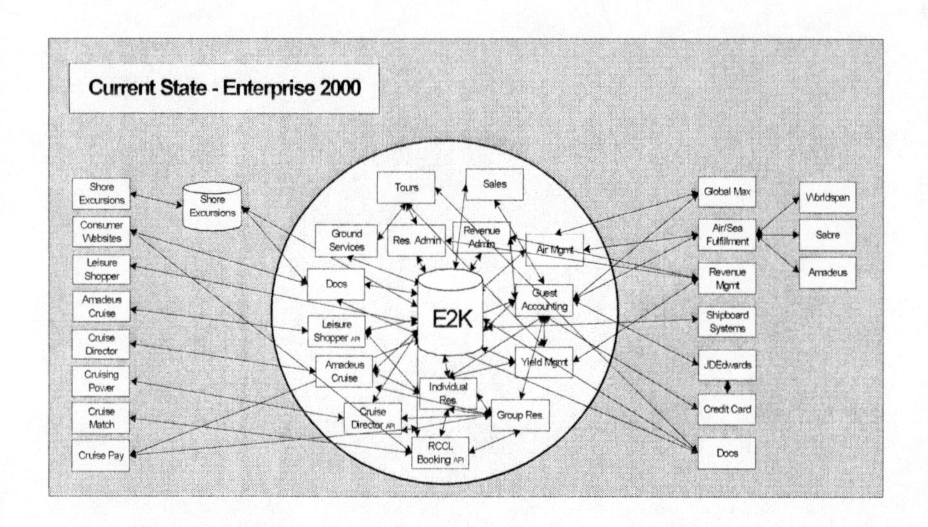

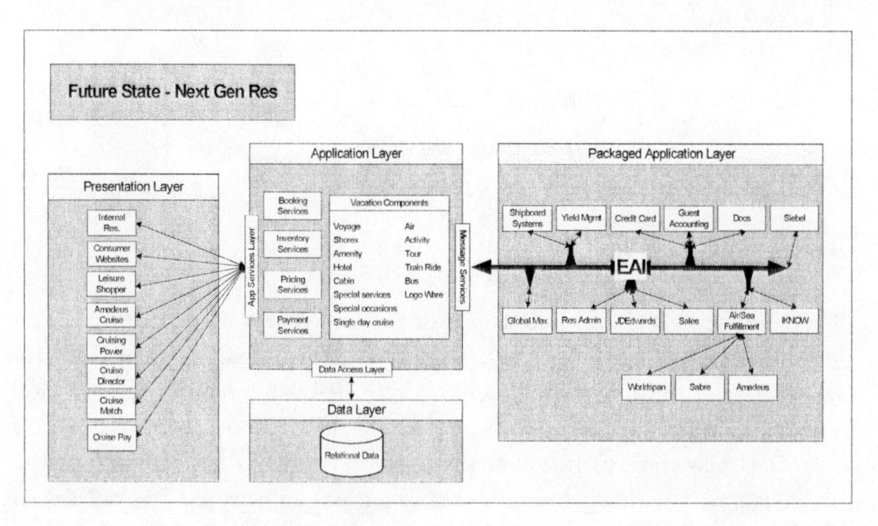

Roadmapping is the best way to reduce large-scale, expensive, cross-departmental projects into small, palatable bites. The first step is to define the business strategy, define the processes of the organization, and align the toolsets in a delivery timeframe with an end goal in mind. Then, the CIO should cut the processes into chunks from a financial and resource perspective. This will allow the company to manage the ebb and flow of the investments as well as manage the training and make changes to the corporate culture over a period of time. The last step is to hold vision sessions that allow for management to imagine what the company will be like once the new tools are implemented. It is important to know where one is going before implementing a roadmap.

Here is a simple example of breaking up large initiatives into multi-year, manageable components of work. The company can choose to increase or decrease spending over time, and the roadmap indicates some degree of dependencies, costs, and time to market:

2003	2004	2005	2006
Imaging & Document Mgmt .2M	Enterprise Foundation Components 5M	Ent. Found. Comp. 3M	Booking, Inventory Mgmt, Other 6M
RCI Web Rewrite .5M *		Booking, Inventory Mgmt - 4M	
Celebrity Rewrite .5M *	Cruise Match On Line 4M	Guest Accounting 7M	External Booking Tools 2M
Shore Excursions 1M *			Ground Services 1.5M
	Flexible Dining .2M *	Packaged Application Integration 2M	Packaged Application Integration 1M
Development Tools 1M	PMS Interfaces 2M	PMS replacement 12M	
		Call Center Apps 3M	Air/Sea Integration 1M
3M	**11M**	**25M**	**18M**

* this is the portion of the total project cost attributed to jumpstart

Process Planning

There is a lack of effective project, change, and process management skills in most companies. Large IT projects will not be successful without project, change, and process management. A CIO should form a group that is focused on process instead of technology. For example, a CIO might hire industrial engineers and operational research people who can go in and do an as-is process map and a to-be process map. Those maps will help the IT and business teams plan how the company wants to move forward. Mapping also gets to the heart of successful technology implementations, which require cultural change management and process management skills. Project management is arguably the most important skill in an IT shop. These projects must incorporate technical skills, development skills, business skills, change management, and process management. It takes very talented individuals to orchestrate all of this into a successful project. Most technology is reasonably straightforward. Failures occur when proper project, change, and process management have not been incorporated.

The Buy-and-Hold Model

There have been many changes in the pharmaceutical industry in the last few years due to lawsuits and a depression in the industry. For example, distribution companies used to utilize a buy-and-hold model to make money. A buy-and-hold model means that a wholesale distributor used to be able to speculate on inventory. For example, a pharmaceutical distributor could buy from a drug company anticipating an increase in the cost of the pill it was buying in bulk. The drug would sit in the warehouse and the distributor would make money on the price appreciation before releasing the pills to the buyer. The distributor could operate to end customers at a loss, because it was making money on the inventory speculation. However, in the past eighteen months, inventory management services agreements have restricted the amount of inventory that can be held, which has essentially eliminated the distribution wholesaler's ability to use a buy-and-hold model.

A CIO in the pharmaceutical distribution industry must now help the company find new ways to make money without using inventory speculation. One way to make money is to move to a fee-for-service model where the distributor agrees to certain service levels for the services it provides both upstream and downstream, and the company is paid for those services. This is not as simple as it may sound, however, as the upstream and downstream customers have never been asked to pay for services they have not put a value on previously. It is very challenging to get customers and suppliers to understand that the company is providing a critical service that should be paid for by them.

Manufacturers enjoy a distribution model that literally delivers their products daily to health systems, alternate care facilities, and pharmacies all over the United States. The end customers enjoy the ability to place an order by 5:00 p.m. local time and receive that product 99.8 percent of the time by 12:00 noon the next day. Yet the distribution companies are considered "non-value-add." This is a perception that must be changed if a fee-for-service model is going to work.

Switching to Fee-for-Service

It is a significant challenge to switch from using a buy-and-hold model to using a fee-for-service model. A buy-and-hold company did not have to focus on customer profitability. The company only had to care about the number of customers, because the number of customers dictated the amount of inventory that could be bought. Therefore, every customer was a good customer. The entire sales organization, in fact the entire business model, was built around customer volume. Of course, the IT infrastructure and tools were built for that as well. In a fee-for-service world where the company is not making money on inventory, customer profitability becomes critical. This requires new tools, new processes, and new organizational structures. New tools and processes to define market segments and profitability by segment and by customer; new organization structure to go to market differently, create different compensation models for sales, and respond to customer interactions differently. Any one of these presents a company with challenges, and to do all of this in a matter of months is unfathomable.

Contract management also becomes a priority. Every single fee-for-service and inventory management agreement has different metrics that have different definitions depending on the manufacturer. Compliance systems need to be introduced, and back-end financial systems must be modified. Individual implementations begin to happen, but all of them need to be linked to process because a company can't disconnect one from another. The switch to fee-for-service has an enormous impact on a company, and the CIO finds the role elevated in relative importance because so much of the company's response requires technology enablement.

Here is an effort to roadmap IT's response to the fee-for-service implications to the business goals of operational excellence, innovation, and ultimately profitability:

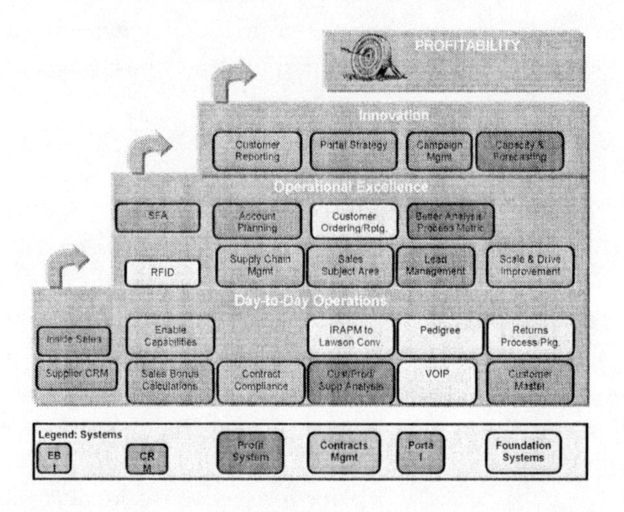

Making Difficult Decisions

The most difficult types of technology decisions a CIO is faced with normally involve relatively mundane technical issues concerning specific technologies. These are usually technologies that have fervent supporters in either or both the business and IT, which become almost religious in nature. Examples may include presentation tools for data warehouse systems, the core database environment for the company, or the type of workflow engine or e-mail to be used. These decisions are generally not of a competitive nature—rather, they are part of the fundamental IT infrastructure, important to the blocking and tackling of IT. Those decisions are difficult to make, because there tends to be such strong opinions about them. At the end of the day, the solutions are generally comparable, but people hold strong beliefs about them. In such cases, the most useful resource for a CIO who has to make technology decisions is trustworthy employees. The CIO should look for people who can be objective in an otherwise very subjective world.

Misconceptions of a CIO

The biggest misconception about making technology decisions as a CIO is that the decision-making process is relatively simple, and that technology itself is relatively simple. While the technology might be simple, the effective application of that technology to the business is quite challenging because of the necessary impact on people and process. It requires a delicate touch to manage business expectations about how simple technology should be. In terms of technology decisions, a misconception is how often the CIO makes the final decision. CIOs should not be involved at such a granular level that they are making technology decisions as much as they are driving the strategies, the linkage to processes, and the business portfolios for where those investments are going to be made.

Three Golden Rules of a CIO

There are three golden rules to making successful technology decisions as a CIO. The first rule is to understand the business so the CIO can add value with technology. The second rule is to ask a lot of questions in order to gather the most information. The third golden rule is that the CIO should hire people who are smarter than him or her.

Thomas H. Murphy is the chief information officer of AmerisourceBergen, a $50 billion wholesale distributor of pharmaceuticals and related health care products and services to the health systems, retail, and alternate care markets. Mr. Murphy is responsible for all information technology activities across five major lines of business, comprised of more than fifteen individual companies doing business as AmerisourceBergen.

He has over twenty years of information technology leadership experience, primarily in the tour and travel industry. He was named one of ComputerWorld*'s 2002 "Premier 100 IT Leaders," and his organizations have been recognized for innovation, resourcefulness, and as being best places to work in the information technology industry.*

Understanding the Business and the Customer

Michael O'Connor
Chief Technology Officer
Novariant Inc.

The Goals and Responsibilities of a CTO

One misconception about the role of the chief technology officer (CTO) is that he or she is always responsible for the information technology (IT) in a company; the role of the CTO is actually company- and industry-specific. For a relatively small technology company, the overall goal of the CTO is usually to drive the technology direction of the company towards success, specifically in terms of the technology that goes into the products or services of the company. Managing what technology is in a product, keeping that technology aligned with available technologies, and prioritizing those technologies based on the product strategy of our company are the responsibilities a CTO must take on in order to accomplish this goal. A good CTO also helps to channel the enthusiasm of his or her technical team into developing products that coincide with the needs of customers. As a result, CTOs focus research and development activities as well as technology activities on the needs of the company's customers. Another responsibility of a CTO is clearly communicating how a specific technology works and the benefits of that technology to the management team.

Making Technology Decisions

In most startup companies, the technology decisions made by a CTO deal with the direction of products over the next three years. In order to make technology decisions, a CTO should participate in all the management meetings and spend time with teams throughout the company, not just the development team. He or she should be aware of the current projects of each team. For example, Novariant has a GPS-based, machine-controlled product that it supplies to the agricultural industry. As a result, I analyze and predict the future needs of the agricultural industry in order to influence the design of new products that will meet those needs.

When making a technology decision, a CTO should utilize three strategies. One is thinking like the customer, whether that is the service technician, the sales representative, or the end user. CTOs should also consider the future when making technology decisions. For example, a CTO should focus not only on what to leave out of the product but also on what to add to the product in the future. These additions will then add additional value for the customer. Lastly, a CTO should remember the benefits of standardization.

Even if it is less efficient from a technology standpoint, it is important to develop a product in a more general way if it ultimately creates more flexibility for the end user.

Technology Expenditures

A technology expenditure can include developing, acquiring, or purchasing a technology. In any case, before making this decision, a company should obtain an assessment of the technical feasibility, as well as the resources needed to bring the technology to the market. For acquiring or outsourcing a technology, a company must assess the technology that exists, the impact of that technology in the market, its level of readiness for the market, and its fit with the existing product line or development team. Based on these criteria, a company can then make a decision about investing in a particular technology.

Working with Other Executives

When a CTO makes a technology decision, he or she works with the chief executive officer and all other members of the management team, because each ultimately has a stake in the final decision. It is critical to communicate and cooperate well with these individuals in order to understand their perspectives. For example, with the vice president of sales and the vice president of marketing, a CTO should recognize that the data is coming from customers. As a result, a CTO who understands the needs of the customers can better cooperate with the vice president of sales and the vice president of marketing. By having an understanding of the entire business and issues associated with each group, CTOs can negotiate and cooperate in a more efficient manner with each of the team members on the executive board.

Hiring the Right People

In many small technology companies, the biggest technology development expense is the employees. Hiring the right people is an essential element of success. A CTO must spend time working with human resources and recruiters to ensure that the appropriate individuals are hired. As a result, spending money on employees is considered a business need rather than a business expense.

There does not seem to be a magic formula for hiring the right people—or at least, if there is, I have not found it. I always strive to surround myself with people who are better than I am, who are creative, and who have a passion for their work and their business. I also prefer to work with people who have the confidence to debate new ideas, but are willing to put their full support behind any decision that is made by the manager or the team for the good of the company.

Setting and Monitoring Goals

Most companies have a well-defined process of setting individual and team goals, which involves formal annual reviews and an annual formal meeting to set and review goals for the upcoming year. In a small company, I find it is actually important to have quarterly meetings in order to review the goals and objectives and adapt them to the changes in the current marketplace. In conjunction with this system, I make sure to have informal meetings with every member of my team—perhaps once or twice a month—to provide and receive "real-time" feedback about company issues, individual issues, and performance issues. I try to participate in project meetings within the company on a somewhat regular basis as well—a luxury of working in a smaller company.

Strategies for Success

While a good CTO must understand technology in intimate detail, the art of being a successful CTO involves focusing on the needs of the customer rather than on the technology. As a result, CTOs must communicate well with customers in order to understand their needs. It is too easy for an executive to remain in his or her office building and rely on second-hand information. There is no substitute for spending time with people in the field—end users, dealers, support staff, sales staff.

Maintaining open communication between the CTO and the rest of the management team is important to understanding the business objectives of the company. Holding weekly meetings with the management team helps to communicate effectively with them. It is the responsibility of the CTO to ensure that the management team understands technologies being implemented and recognizes the benefits and limitations of those technologies, as well as alternate technologies.

The alignment of the needs of the customer and the objectives of the company is the critical factor for success. A CTO must also be familiar with existing technologies and the markets targeted by a company in order to implement his or her products successfully.

For a company to continue its success, a CTO must spend time thinking about the future of the market and the future of technology in order to prepare the company for future changes.

Overcoming Challenges

One challenge of the CTO is making time to focus on the strategic issues of the business. To overcome this challenge, CTOs should surround themselves with self-motivated, extremely capable team members. If you know that the engineering, research, and operational aspects of the business are running smoothly, it becomes easier to focus on strategy.

The most difficult types of technology decisions involve prioritization. The challenge is choosing which technologies to pursue and aligning the resources to acquire them. It is important to recognize that an agile company cannot implement every technology that is identified as useful. By thoroughly assessing a company's business needs and identifying a limited but key set of existing and new technologies required to meet those needs, a CTO and his or her staff can overcome this challenge. The projects that bring the most value to the company by benefiting the customer will also bring success to the company.

Return on Investment

In a startup product company, it is imperative to predict and measure the return on investment for a new technology. Depending on the technology, this return may be based on increased sales due to increased market share or increased market penetration. It may also be based on expense or cost reductions. In either case, predicting this impact—especially a sales impact—can be quite challenging. Measurement, however, is usually straightforward, and this step is critical to provide feedback that will improve the prediction process.

While a graduate student at Stanford University, Dr. Michael O'Connor led the team that invented the first ever farm tractor steering control system using GPS. The project, sponsored by John Deere, ultimately produced four Ph.D. dissertations and was the seed for the AutoFarm product line.

Dr. O'Connor obtained his B.S. from M.I.T. in 1992, his M.S. in 1993, and his Ph.D. in 1997 from Stanford University. He was recently named to the 2003 list of the world's "100 Top Young Innovators" by Technology Review *magazine.*

A Balanced Approach: Technology and Business in the Role of a CTO

Bob Kruger

Chief Technology Officer

NDCHealth Corporation

Defining Success as a CTO

The goals of a chief technology officer (CTO) at NDCHealth may be unique, because the CTO faces two distinct types of customers: those who purchase our products and those internal ones who use our services. For the external customer, the main objective is to increase the level and pace of product innovation. That needs to be a long-term strategic effort, not just a short-term tactical endeavor. For the company, the CTO must bring along the various teams scattered throughout the company in order to further company objectives. This internally facing role is particularly difficult, because the role of the CTO is an indirect yet influential position. The challenge is two-pronged: how to move forward as a single company rather than a collection of assets and how to bring about that convergence without necessarily having everyone report to the CTO. For example, the CTO may control resources by virtue of signing off on the overall budget, but each individual business unit in the company really drives its individual budget to achieve specific business unit goals. These objectives are not necessarily in alignment, further complicating the CTO's ability to succeed. Therefore, a successful CTO is able to conceive things as an individual but work with various company organizations so that the final product is a team endeavor.

Going back to the customer-facing role, the CTO must be aware of what is going on in the industry as well as within the company. The successful CTO is smart, but also practical, from a technology standpoint. The person must be able to think strategically and see an idea through to a long-term conclusion. It is also essential to connect technology to the business. It is equally important to be creative and able to recognize good ideas and capitalize on them. The best technology or the most elegant solution is not necessarily the thing that results in the best product or the greatest revenue potential for the company. Making a decision just for the sake of technology is not the answer.

Understanding the Technology

The CTO helps the company make technology decisions by forcing the appropriate reviews and conversations. For example, the need to improve a product's subsystem for editing online transactions can be handled by adding people, incorporating a purchased rules engine (i.e., software that enables the creation and execution of business logic, or rules, for the purpose of ensuring the correct entry of data or making workflow decisions), or building a rules engine. If the company already has a rules engine in use within another product, why not start there and look at how it might be improved for both products' benefit?

It is also helpful to make sure that people involved in making technology decisions are actively engaged with other companies and industry partners. It is important that they are visiting other companies, going to conferences, reading about and trying new technologies, and staying current with technology trends in order to make the best decisions. Also, it is vital to understand the company's business and its requirements on technology in order to deliver the best possible solution for customers.

An equally important strategy to deciding about technology is having a consistent concept about a product's usage from the beginning stages of creation through to marketing the product. It is not enough to explain the idea but balance that with the practicality of it. A CTO must highlight what is working, what is not working, and create an action plan to solve the issues created.

Adding Financial Value

One of the biggest ways to have a financial impact on the company is to help people see which products need to be obsoleted so resources can be redirected to products that really have the potential to move the company forward. Even companies that have been bringing in strong revenue can stagnate or have slower-than-potential growth because they're too fixated on what's already being delivered. A CTO needs to anticipate the "innovator's dilemma" (as coined by the book of this same name by Christian Clayton) and plan for obsolescence. The alternative is that the competition delivers a disruption and obsoletes their business for them. (Anybody thinking about creating a superior videotape today now that DVDs have taken hold of consumers?)

Another way to financially affect a company is to create and leverage technologies that can be shared across the business units instead of being used within one business unit. Ultimately, that would save time and money.

For example, if it is possible to use the same rules engine on four products instead of one, development overhead can be reduced dramatically. In fact, with greater commonality of components, the underlying infrastructure can be transformed into a "platform" for higher levels of value that can be spread across more products. For example, a claims management system that has a common foundation for hospitals, clinics, labs, and physician offices can be a great base to host other value like a business analysis product. The fact that there exists a broader segmentation of customers makes the potential revenue more enticing to the solution supplier and may therefore incentivize the supplier to support the underlying claims management product.

Working with the Team

There are many members of the executive team that a CTO has to work with in order for a company to be successful. After all, from an internal IT perspective, the CTO owns responsibility for enabling the necessary systems to support the company's business. On the product side, it is important to work well with the heads of development, architecture, product management, and the business unit general managers. Essentially, a CTO has got to buy into strategies and tactics promoted in various areas of the company, and this necessitates a good understanding of company and department objectives and operations. This dialogue occurs so that when technology decisions need to be made, the various departments can understand the rationale and readily execute on the results.

In order to work within a team, it is best to set well-estimated strawman goals and timeframes, and then let the individuals within the team determine how best to reach those goals or recommend solid alternatives. Furthermore, it is a best practice to ensure that team members are cognizant of target market windows or other industry events to maximally stimulate their motivation. The goals and results should be reviewed on a regular basis with a minimum of quarterly reviews and a maximum of weekly reviews, depending on the state of the project and its criticality to the company.

A customer-focused team will always fixate on the customer first and the company second. Technology should not be the driver of a company but rather an enabler to fix external or internal customer problems.

Making Decisions

A good CTO can help the executive team understand what innovation means and how to look at it in terms of disruptive and sustaining types of changes. For example, is it the right approach for a company to incrementally enhance a product or spend more dollars on an entirely new approach or adjacent product? Helping people to candidly converse about those things in order to make good decisions is essentially, and a frame of reference for such discussion can be driven by the CTO. In many instances, part of the problem will be that executives are unwilling to make tough calls without all possible information, something that is rarely possible. Being able to bring people together to consider these tough decisions and helping them to understand what's important and what's not is an essential technology management function.

Change is Expensive

Concerning internal IT operations, telecom costs comprise some of the biggest routine expenses for many companies. For a company like NDCHealth, whose business is substantially built atop a health care transaction environment, this is certainly the case. Often, implementation of an enterprise resource-planning product presents significant costs, many of which span multiple years. Ongoing hardware maintenance and upgrading can be costly.

From an external customer-facing products perspective, the money goes more to people than anything else. Given that a company's value is tied up in the intellectual property delivered by people, this makes a lot of sense.

By the way, there are many ways to perform due diligence on potential technology expenditures. Sometimes, it's as simple as meeting with vendors, researching the products, and deciding whether to purchase. In other instances, particularly in the acquisition of technology to embed in products or for very significant internal IT investments, it may be necessary to hire relevant people to come in and do appropriate due diligence or leverage smart people from other parts of the company (who don't work in an area similar to what's being acquired). That way, a deep understanding of the technology and its planned evolution can be obtained in a way that won't contaminate the company should the potential acquisition not be consummated.

Picking a Platform

From the standpoint of external, customer-facing products, the most difficult decision is still which platform to chose. The differences between Windows, Linux, and Java still require considerable examination. Marketing must be discussed. Delivery and usage methodologies must be investigated. The target customers' perception of the various platforms' security must be examined, and so on. These are decisions that need to be considered very early in a product's lifecycle.

Internally, a company needs to analyze a platform's strengths and weaknesses in the context of its planned usage. Whether the company has an existing base of knowledge about the platform is also important. If internal systems will also be used to support customer systems, that might additionally factor into deciding which platform is best. Fortunately, some of these particular considerations are becoming less of an issue as the use of Web services increases and shields customers from the nuances of underlying platforms.

However, not every decision can be based just on technology. The best technology does not always represent the winning solution. Having something that's architecturally pure or elegant from a technology standpoint but disconnected from the business or the company's overall strategic direction is just not going to work. It has to be aligned. There's got to be balance and synergy.

Calculating Return on Investment

It is easy to calculate return on investment for products if a company has comprehensive business plans. A company should produce a semi-business plan for internal products. Return on technology investments is still based on business plans. As a company acquires new technology, it should always look to see what the return timeframe is and whether the technology can underpin or help to yield multiple products. Products should be examined based on whether or not they are providing the necessary customer value and horsepower in terms of performance to enable future growth. A worthwhile technology that saves the company money, makes people more productive, enables the company to perform better, and allows the company to be more responsive to its customers will always be a winner.

Bob Kruger is the chief technology officer of NDCHealth Corporation, a more than $385 million public health care information technology company serving all sectors of health care with information, transaction, and software product solutions. He reports to the NDCHealth chief executive officer as a Section 16 officer in charge of technology strategy and vision, for customer and internal information technology solutions.

Most recently, Mr. Kruger was senior vice president of product development and chief technology officer of Citrix Systems Inc., a more than $700 million global public enterprise software company. As the technology leader providing the software products for secure access to applications, information, services, and people using any device over any network at any time, Mr. Kruger demonstrated strong technology and organizational skills, along with keen leadership, creativity, and business and common sense.

Mr. Kruger has over twenty-seven years of technology career experience. Prior to joining Citrix, he held key positions at BMC Software, including vice president and general manager for e-business solutions, where he defined, built, marketed, and supported leading-edge service level management solutions on distributed systems. As during other times in his career, he was charged with turning around languishing products, conceiving and delivering new innovation, and improving operational efficiency. Through his efforts, BMC dramatically increased its portfolio of Windows and e-business solutions, and won awards and praise for its products from industry trade publications and analysts.

Prior to BMC, Mr. Kruger spent more than ten years with Microsoft, most recently as general manager for systems management products in the server applications division, where he developed, marketed, licensed, and supported Microsoft products designed for the management of Windows systems. He was instrumental in establishing the computing industry's Web-based enterprise management standard and the crucial underlying technology, in addition to collaborating with Bill Gates on his second book, Business @

the Speed of Thought. *Throughout his years with Microsoft, Mr. Kruger had successfully taken charge of challenging corporate relationships, the marketing of new technologies, oversight of key technology standards, and lobbying with United States and foreign government agencies, and has been responsible for networking and UNIX-based businesses.*

Mr. Kruger is a graduate of UCLA with additional studies at the National Autonomous University of Mexico and the AeA/Stanford Executive Institute for Management of High-Technology Companies. He speaks Spanish, Portuguese, and some French. He presently serves on the national board of directors for the AeA (the largest high-tech trade association in the United States) and the industry advisory board of the Florida Atlantic University engineering and computer science department. He has sponsored the Victory Living Programs organization for guidance and training of people with developmental disabilities or traumatic brain injuries.

Getting in Tune with the Business

Basil Kanno

Chief Information Officer

PowerNet Global Communications

My Primary Goals

The primary goal of a chief information officer (CIO) is to create an organization with a vision that is solely focused on the best ways of supporting the various aspects of the business; now and for the future. The CIO must always keep in mind that their purpose is to make things efficient, which translates to cost savings, which yields greater profits.

One key factor to a CIO's success is getting a detailed understanding of the business and business processes across all organizations within the company. Once a technology team understands the business in these ways, it can implement solutions to any business issues or needs that arise. Another key factor to a CIO's success is strong leadership skills. A CIO must be able to assemble and manage a team that is dynamic and not afraid of change, committed to his or her vision, and aligned with the overall company direction.

The Art of Being a Technology Leader

Typically, CIOs have been viewed as technology jocks. It is important for a CIO to understand business objectives and their requirements. In fact, a technology leader should be just as business-minded as technology-minded.

The art of being a technology leader also involves building relationships with peers, other C-level executives, and other organizations within the company. This is very important, as you must realize that you cannot get things accomplished on your own and in a silo. You are going to need the expertise of other organizations to accomplish your tasks. By building strong and effective relationships, others are more willing to work with you and help you achieve your goals.

Third, a technology leader must be an active listener. Listening is grossly taken for granted. We all think we are good listeners. Listening is not merely hearing the words that are coming out of the other person's mouth. It is about trying to understand what the other person is trying to communicate to you. That can be achieved by probing, asking questions, using visual illustrations for clarification purposes, and so on. When you actively listen, you will establish an environment of trust, respect, and creativity between you and your peers, your team, and so on.

Lastly, a successful and effective technology leader needs to have common sense. In many situations, executives throw everything they've got in their arsenal into solving a problem that requires a simple, logical solution. By using common sense, a CIO can pinpoint these logical solutions more efficiently and effectively.

Working Well with Others

When dealing with their peers, technology leaders should strive to understand the operations of their peers' organizations and the daily challenges that face them. With that understanding, technology leaders must then view the world from their peers' perspectives. The other executives should also understand the operations, challenges, and perspectives of the CIO. Acknowledging this variety of perspectives allows the executives to avoid potential conflicts and to maintain a sense of cooperation.

Team Roles and Responsibilities

A CIO can only be as successful as his or her team. One of the main responsibilities of a CIO is building a long lasting, effective, and efficient technology team. The CIO must realize and be comfortable with surrounding themselves with talented individual—individuals who are driven, ambitious, and unselfish. But the key point is that when these individuals are put together, they must operate as one. There is no room for ego in a team, especially one built for longevity.

The CIO must be confident in his or her team. He or she must empower and give appropriate levels of authority to each member of the team. The CIO must also provide guidance and feedback when appropriate.

In turn, the main responsibility of the team is to execute on the vision established by the CIO in a cost-effective and timely manner. The team is also responsible for giving the CIO constant feedback from the field and marketplace, since a CIO's vision and decision-making is largely dependant on that.

Setting Goals for the Team

It is important for a CIO to set goals for his or her team. Goal setting gives the team a sense of direction and a clear picture of what needs to be done and when it needs to be done by.

The goals a CIO sets for his or her team should be aggressive but realistic and achievable. There is nothing worse than setting unrealistic goals, as this kills the team's morale and motivation, which is detrimental to the CIO and the company. Goal setting should be an exercise carried out between the CIO and his or her team in partnership. The team must understand what the CIO is trying to achieve, and the CIO must listen to his or her team to get a realistic understanding of what the team is up against.

It is also crucial to perform a "lessons learned" session when the project or initiative is all said and done with. This exercise will help the CIO and his or her team enhance their goal-setting skills.

Choosing an Approach

When creating methodologies and processes to streamline the operations, a technology leader must ensure that they are in the simplest form possible without sacrificing their inherent benefits. Oversized and complex methodologies and processes are nothing more than overhead, which slows an organization down tremendously. A CIO must always encourage his or her team to think simple first and build from there. Unfortunately, there is no "one size fits all" concept when it comes to methodologies and processes. Therefore, when creating these practices for a particular organization, a CIO must ensure that his or her team is taking the organization's capabilities, culture, and size into consideration.

In a typical corporate environment, the number of initiatives/projects surpasses the limit that the company's support functions can handle. This can easily bring a company to its knees from a productivity perspective and can be quite costly. Therefore, CIOs must put into place a mechanism that can bring sanity to that environment. This mechanism should prioritize these initiatives/projects based on the effort involved in implementing and supporting them and the value they yield. This mechanism will also ensure that the CIO's team is always working on the right thing versus just busy work, which brings no value to a company.

Identifying and Overcoming Challenges

Technology has a very short shelf life. One challenge is remaining abreast of the constant changes in technology. Not doing so can be detrimental to a CIO and his or her company. It could result in the company losing its competitive edge in the marketplace. It could result in lost opportunities in reducing operational expenses. A CIO can keep up with these frequent changes by dedicating the necessary time. With that time, the CIO must read relevant technology articles, attend technology seminars and tradeshows, talk to industry experts, and so on.

Another challenge is making difficult technology decisions based on short-term versus long-term goals. My choice is often to think long term and then find a reasonable solution for the short term, if one is available. If there is no such solution, then sometimes a technology leader is forced to make the short-term decision.

A third challenge for small to midsized companies is not having the leverage that larger companies have when dealing with external entities like partners and vendors. This spans the whole spectrum from negotiating contract terms to pricing. So, the challenge for a CIO is how to get the best possible deal without having the big name (e.g., Fortune 500 companies) to fall back on. Most partners and vendors these days prefer to establish long-term relationships with their clients, as this can yield greater sales opportunities over a long period of time. As such, the CIO of a smaller to mid-sized company needs to educate his or her partners and vendors on his or her long-term vision. This process should point out all the potential opportunities for these partners and vendors. When partners and vendors see and understand the long-term potential, more times than not they will be more willing to make certain concessions on contract terms, pricing, and so on.

Expenses

Reducing expenses is not something you do once in a while when budgets are tight or revenues are down. It is a practice that always needs to be on the forefront. It needs to be part of an organization's fabric. Expense reductions can come from various areas, but one of the main areas is business process reengineering/automation. Typically, business process reengineering/automation results in the greatest expense reductions. But a CIO must be careful not to get caught in the trap of automation for the sake of automation. You must automate what makes sense, with what makes sense, when it makes sense. Sometimes, keeping processes manual is the smart approach. For example, if it costs you $100,000 to automate a process, which yields a $1,000 of expense reduction per month, then this is a process best left alone in favor of other expense-reducing opportunities.

Return on Investment

Return on investment needs to be part of a CIO's everyday philosophy. A return on investment analysis must be conducted on any initiative/project requiring any type of resources. CIOs must ensure that their team is working on initiatives/projects that yield the highest returns first.

Our Changing Role

Technology leaders are starting to be more in tune with the business. Previously, there were CIOs that never took the time to understand the various aspects of the business. However, without understanding the business, a CIO cannot make an effective technology decision to support the business.

Basil Kanno has over twelve years of comprehensive information systems experience with expertise across a wide range of systems and applications. At PowerNet Global, he ensures that all facets of the business, from process reengineering to product rollouts, are properly supported by the internal systems. Prior to joining the company in 1999, he was a senior consultant with Ernst & Young as part of their telecommunications practice. Previous to that, he was at EXL Information Corp., where he managed strategic systems development for their flagship telecommunications billing product.

Mr. Kanno joined PNG as chief information officer in 1999. As the chief information officer, he was a key contributor to the tenfold growth of the business. Mr. Kanno earned his bachelor's of applied science degree in computer science from Simon Fraser University in Vancouver, British Columbia.

Providing the Company with the Tools Necessary for Growth

Jeff Frederick

Chief Technology Officer

Excel Group

Providing Tools for Everyone

I work for everyone else here; I provide the tools necessary for everyone to do their jobs. I look into the future, as far as research and development are concerned, to see what's coming down the pike that might be a better tool for us to use. I then implement those tools and make sure they're up and running for everybody. Just like everyone else, we're looking at dollars and cents and trying to make sure we don't break the bank while providing people what they need to do their jobs. We want to make sure it's all done at a reasonable cost.

I've worked just about every job here, and I understand what people are going through to get their work done. It's all just done on a bigger basis at this point. Being able to see what each person needs has helped me to make good decisions for the company, and it has helped us grow from practically nothing to a $10 million business.

Making Technology Decisions

I'm the one who makes the technology decisions for the company. Our business is made up of four divisions. It all started with a courier operation, so we learned a lot about transportation. That got us into trucking, which got us into warehousing, which got us into archiving, which got us into imaging. We have four divisions: Excel Courier, Excel Transportation, Excel Archives, and Excel Imaging. As the chief technology officer (CTO), I'm in charge of all four. I'm soon getting an assistant for the first time, on a part-time basis; I'm looking forward to getting that person in to help with the day-to-day work I can't anticipate. That will allow me to stay more focused on the major projects we have going.

The general managers who run the other companies come to me for business tools, software, hardware, or whatever they might need, and then we budget those appropriately and get them the equipment or software in a timely fashion.

I manage everything from the communications system to the phone system, the security system, the video capturing system, all the servers, all the computers, all the pagers, and all the cell phones. I'm also in charge of a lot of the facilities, the HVAC, and things like that. I've been here for seventeen or eighteen years now, and I'm now in charge of most of the equipment and communication areas.

We're working harder every day to keep within the budget and to project our needs for the next eighteen months on an ongoing basis. That's an important part of my job: making sure we know where our money is coming from and what people will need in the future. We're always trying to monitor those things to make sure we're not running into any big surprises. I constantly review software and our vendors to make sure that: (1) we're getting the best software available for our team, and (2) that it's financially feasible to stay with that particular service provider. Those are the main things I'm trying to accomplish.

Understand the Decisions

My strategies are to make sure I precisely understand the end users' goals before throwing time and money at a decision. I try to consider every option. If there are several different vendors who offer products, make sure you look at each one. If it's a big decision, you want to get references to make sure other people have been treated correctly by the software or hardware vendor, whichever it may be, and then follow through and make sure you're staying up to date with the latest releases they have. Make sure you've got the support you need to keep the operation going smoothly.

Once you understand what your goals are and what you're trying to accomplish with your technology decisions, it's just a matter of making sure the products you're buying are going to work best for the end user and for the company as a whole. It's not magic or anything like that. It's about having a clear understanding of what you're trying to accomplish. Once you know that, your decisions become a lot easier. Each decision is different, so you have to be certain of your customers' and your employees' needs before you start buying things and changing technology.

Setting Priorities

My biggest challenge is balancing all the different aspects of the job. The managers at Excel Group sit down and talk about what we're going to need in the future. We also have weekly and quarterly management meetings to go over in more detail all of the things we've done. We review budgets and actual numbers and how all of them break down, and my managers keep me informed of what they might need in the future. We're trying to focus on large projects and get them completed one at a time; we're trying to set priorities to make sure we're addressing what needs to be done in the order of its importance.

Working as a Team

I work most closely with the chief executive officer, and we're trying to work together more closely every day to be certain we're on the same page. We want to understand our priorities, focus on the ones that are most important, and complete them. We're meeting every Monday for about fifteen minutes to talk about our weeks; this way, I can see if he needs anything special or if I need anything special, and we can make sure we're on the same page as far as getting projects finished.

To set goals for our team, we're listing tasks on what we call our "one-pagers"; the projects we know are coming up, initiatives in the future, and so forth. We're going through those tasks and changing priorities until we finish a particular project. Then we check that off and move on to the next one. This requires dealing with the chief executive officer and the general managers of each of the companies to make sure we're meeting their needs as far as technology goes.

I don't give a whole lot of advice to other people. It all goes back to the general managers and employees making requests while keeping our goals in perspective. Sometimes, people ask me to focus on something that doesn't really fit into what we're trying to do at the moment, or they want me to look at a particular kind of software and I have to explain to them why we use something else. I anticipate being more involved in that as I have more people working underneath me. Right now, however, my main goal is to make sure I'm giving the customers, officers, and employees what they need.

Ensuring Success

We spend so much time together; if the management team doesn't understand our objectives, we have a bad communication problem. The people at Excel understand that my part is to make sure everybody has the software, hardware, communication equipment, and whatever technology they need to do their jobs successfully. The managers will let me know if we do not have what we need, and I'll make sure that if it's something our customers and/or employees need, we will get it.

Don't break the bank by going out and getting cutting-edge things just to have them. Make sure it fits into your business plan, and make sure the tool you're getting is going to solve problems for you. Make sure there's some sort of payoff on the investment—there's no reason to waste money.

Expenses and Return on Investment

The biggest technology expenses we incur come from providing all of the software and hardware. We also want to put a generator in our building, which will cost a lot of money. I've had that on my budget for the last couple of years, and I would sleep better if I knew our building was serviced all the time by a generator. Up to now, we've been reluctant to make such a large purchase. At some point soon, it's going to have to be done. We've been very fortunate with our power supply, but I always worry about not having electricity. I have small generators that can keep my servers, phones, and some workstations up and running. We can keep our business going, but that super-generator is something I'd really like to have.

I sit down weekly with the chief financial officer, the chief executive officer, and the general managers, and we decide when we're going to pull the plug on something or start working with it. For instance, we're thinking about changing the software that runs our archive business. Our current software does well, but it doesn't have a good Web interface. That has forced us to look into some other options. We've gone through about eight hours of demonstrations so far with another company, and we are going to visit them and their largest client next week—we spend a lot of time and effort making sure something is going to work better for us before we even think about changing. It would take a lot of work to change software from one vendor to another; if we're not positive it's going to be better for us in the end, it's not worth it. At this time, we're moving more into the financial end of the decision to make sure it's going to be priced out in a way that will be beneficial for us. We know the new company's Web module is much better, so we're starting to get the references we need and we're going through and making sure this product is exactly what we need to improve our business.

We've gotten a little bit better as a company at saying, "This is what we budgeted and this is what we actually spent." I'm able to run the information technology (IT) side of our operation for a lot less than most other companies by deferring things that don't have to be done and purchasing the technology, hardware, and reliable equipment that will give us the most return on our investment. You have to be smart enough to discern when it's time for change or when it's time to move on. We don't have a perfect calculation for this, but so far I've gotten a great return on investment for anything we put into the business. Another challenge is trying not to be cheap about things and knowing when to spend money on something. We have a bit more money to spend on things these days, and we don't have to be quite as frugal as we did when we started. If something is working well, the company's growing, and the employees are efficient, then we're getting a good return on investment.

The Worth of Technology Investments

If a technology investment makes it easier to get clients from the marketing side, then it's something we'll definitely want to get in the door. An example is our eTrac system, which allows our customers to place orders online; they can track everything, run reports, and so forth, and we have gotten a lot of clients because of this service.

Informally, I'm constantly evaluating everything we use and making sure it's the best thing we can possibly have for our employees, for our business as a whole, and for our customers.

Utilizing Technology Resources

The Internet is so powerful—there's so much information and so much research you can do. That's always where I'll start when researching technology, and then I'll move to the associations and so forth. I meet a lot of people who might be strong in one area or another, and I'm able to network with them and ask them what they've done in the past. Then I've got several people I've worked for, or who have worked with me in the past, and I'll talk to them. I've got a good friend who owns a consulting business, and I ask him questions about things—he's always got great answers for me if I'm ever confused about something. He taught me a lot when I first got into the business, and he is still helping me today. I also have a very close friend who helps me with the integration of systems and high-level computer security, which is extremely helpful.

The Changing Definition of a CTO

I hadn't heard the term CTO until a few years ago, so I believe the role has just been recently defined. Everyone now realizes that technology and data are the most valuable assets of a company, besides the actual people. The chief financial officer, being in charge of finances, is very important. The chief executive officer, being in charge overall, is important. Technology has worked its way up there, too. It didn't used to be quite as important as it is today; without the technology and the CTO, a company would most likely go out of business.

CTOs need to understand the company for which they're working. They need to understand the philosophy and the core of business to be able to make successful technology decisions. It's very difficult to work for a place if you don't understand what the front-line people need; to be able to make great decisions for them, you need to sit down and talk with people working all aspects of the job to make sure they have the tools they need to do their jobs successfully. Then you need to back all that up and make sure you can maintain it, that you're not going to lose the data, and so forth. It's about trying to keep everything safe and trying to keep the wheels rolling. CTOs have to try and look into the future to make sure they're making the best decisions. You also have to be able to change direction if you realize you haven't made the best decision.

Know your project and understand down to the end user what needs to be accomplished. Don't spend more money than you need to. Don't go out buying cutting-edge stuff if it's not going to grow the business and have a good return on investment. Make sure you're not breaking the bank, and make sure you're getting your people what they need to do their jobs.

Jeff Frederick was born in Belleville, Illinois, in 1962 and moved to Vienna, Virginia in 1965. He graduated from Oakton High School in 1980, and from James Madison University in 1984 with a degree in management information systems. Mr. Frederick worked as a consultant for a few years until becoming vice president for Excel in 1988.

Dedication: *To Donna McGrath.*

Adding Value and Solving Problems

Tom Bishop

Vice President and Chief Technology Officer

BMC Software

Expectations Surrounding the CTO

In those businesses that require a chief technology officer (CTO), the expectation is that much of the strategy, direction, and vision of the company require multiple perspectives, one of them being technology-oriented. Because technology evolves quickly, it becomes imperative to have someone whose job focus revolves solely around how technology may be applied either offensively or defensively in order to ensure that the best value is delivered to customers on an ongoing basis.

Technology Decisions for the CTO

Because the technology industry is a vibrant entity in today's business world, the specific list of new technologies or inventions available changes continually. For this reason, the CTO must be prepared to work with others in the company to facilitate the decision process for a number of questions about how to apply today's technology to the company's products and services.

Software used by large companies to manage technology-based assets is known as "enterprise management software." While most PC-based computer technology is widely considered to be mostly self-contained, enterprise management software differs in that its architecture may be described as a three-tiered model. The first tier consists of the environment being managed by the software; a local presence is often required to reside within that environment in order to provide the enterprise management tools with the interface(s) they need to do their job. This local presence is widely referred to as "agents," but other terms are also used. Agent technology is a highly debated subject at this point in time, with the chief debate revolving around whether or not agents are really necessary and how light they may be built and yet still be functional.

The second tier is typically comprised of the management server (or servers), which is the metaphorical "home base" to which all agents report. A number of discussions exist revolving around what types of technologies would be most helpful to ensure that the management server is simple to install, configure, and maintain, but regardless of how they're built, the management server configuration is really "the brains of the outfit," and it's this tier where most of the engineering effort is applied during development.

The third tier of enterprise management software is the management client or "user interface" that delivers the information captured and stored by the management server(s) to various constituents or stakeholders within the information technology (IT) organization. The client software is typically used both by individuals in the IT environment and increasingly by individuals within the lines of business being served by the IT organization. In almost all cases, these individuals typically wish to have some view into the collection of information being collected by the management server.

The CTO often becomes involved in discussions regarding how best to build each of these three components. For example, in the case of the client or user interface, should the solution use a "fat" client (i.e., a large piece of code that must be installed on the desktops or handheld devices of any individual wishing to view that information), or should there be a "thin" client? Fat clients typically provide a greater opportunity to offer rich client displays and functionality, but at the cost of a distinct third component that must be developed, distributed, installed, managed, and maintained. Thin clients have traditionally had to limit the richness of their displays and functionality, but require almost no installation, management, or maintenance, thus reducing total cost of ownership of the overall enterprise management solution. The CTO must often help determine which tradeoffs are appropriate in each case and then make the necessary decision between the two basic alternatives.

Additionally, the CTO may make decisions with respect to open source, an increasingly attractive development or technology option for companies whose business is to build and design software. Opportunities often exist to use open source components in the development of a product regardless of whether a company's products and components are ultimately distributed and supported in open source form.

Finally, the CTO often becomes intimately involved in both the use and development of standards in a number of different areas. These standards play a key role in how a product may be designed and developed.

One Potential Upcoming Change in Technology

The emerging area of computer-based business process and business process engineering is poised for tremendous potential growth in the next several years. While the Internet has been available for some time, because of its relative fragility and difficulty of navigation, it did not become ubiquitous or widely used until the invention of the World Wide Web and the now-ubiquitous Web browser. With the World Wide Web came an ease of Internet use that facilitated an unprecedented number of business opportunities (with companies like eBay, Yahoo!, and Google being just the tip of the iceberg). The potential for a similar business revolution exists with the evolution and increasing use of Web services, which in combination with Web-based business process and business process engineering technologies could usher in as dramatic a change to the business climate (in the form of agile business process reengineering) as the World Wide Web before it.

Qualities of the Successful CTO

To be successful as CTO of a company, the individual must have a clear idea of what the company is striving to accomplish in the form of strategy, business objectives, business plans, and market goals. When making decisions around technology and its use both within the organization and within the company's products and services, it is imperative to remain attentive to the business objectives, market objectives, and customer feedback at all times. The CTO may spend a significant amount of time interacting with customers or speaking with analysts and the press in order to better understand these various components of the business.

In addition, the CTO must ensure that decisions are made in a business context. This means a number of different factors (such as cost, time to market, and maturity of the technology) must be taken into consideration before making any decision, even those that are technology-related. For this reason, it is imperative that the CTO establish good working relationships with the other individuals in his or her own company as well as any other organizations with which the company will be doing business. In this way, the CTO may ensure that many different perspectives have been taken into consideration before a decision is made.

The CTO works with a number of different individuals within the company, including sales, marketing, strategy, legal, and finance, as well account managers, salespeople, and a variety of organizations. It is critical for the CTO to build and maintain as many positive, constructive relationships as possible both within and outside the company itself in order to maximize the company's potential for success going forward. The CTO must always recognize that a company is comprised of individuals with different perspectives, goals, and agendas, all of which may be valid when viewed from each of these different perspectives. To ensure that any decisions made are in the best interest of the company, it is critical to remember that these individuals have valid information that should be taken into consideration before the final decision is made.

Maintaining an Edge

In order to ensure that a company remains competitive, the CTO may have individuals responsible for a wide variety of activities. For example, one person may be responsible for the company's standards activities. In addition, there may be a regular trend monitoring function in which the technologies coming into or out of fashion are routinely examined against a well-known set of criteria. A governance board may be put in place, providing a regular forum for the discussion of relevant issues of the day. The governance board also provides various company stakeholders across the company an opportunity to participate in discussions about the decisions that will ultimately affect them and the company itself.

Advice to Live By

It is critical to keep in mind that the goal of any company is to effectively serve its customers; without them, the company loses its ability to earn revenue and thus ceases to exist. Therefore, every effort must be made to provide a product or service that is relevant and valuable to the customer.

In addition, a company should always take into consideration the timing of its products and marketing strategies. For example, there had been repeated attempts to introduce the microwave oven to the market since the 1930s. However, it did not succeed until the 1960s, because at that point women were entering the workforce in substantial numbers and a device that could provide a hot meal in a short amount of time became valuable. It is critical for individuals working in technology-related fields to remember that success is reliant on market timing. As someone once said, "Timing is everything."

Strategies for Success

To be successful as a CTO, it is first critical to keep in mind that technology must have some value and solve some type of problem in the real world. Secondly, the CTO must take market timing into consideration before making a significant investment in a new technology.

Additionally, it is crucial to keep in mind that technology decisions are often multifaceted; business, market, and legal constraints must be taken into consideration in order to make the best technology-related decisions possible. Finally, the CTO must consider whether a technology can be adopted incrementally or whether it will be necessary to replace an entire existing technology.

Significant Expenses

Typically, an IT team will expend a great deal of resources looking at, reading about, and talking to people about new technology. A fair amount of time is spent on the acquisition of technology, which means that in most cases an acquisition budget will exist within the company for that purpose. The innovations needed to perpetuate the company's mission and existence may often be funded by other businesses more willing to take some risk from an innovative perspective; when the new technologies are no longer a risk, the company will then intervene and either buy the technology or the business itself.

Helpful Resources

Trade shows and other industry events may be helpful to the CTO in providing an overview of what new technologies may prove popular or useful in the near future. In addition, trade magazines can be helpful, though it is necessary to ensure that the CTO remains cognizant of the fact that the magazine may not be entirely objective about the technologies being discussed. Trade books tend to become outdated quickly, while the trade press, the Internet, and other types of publications are typically better able to remain current with the technologies on an ongoing basis. Finally, the CTO should always speak with individuals within the company as well as customers about the technologies being utilized and evaluated.

The Changing Role of the CTO

The role of the CTO has changed dramatically in the past few years with respect to technology decisions. A decade ago, there was more willingness in the industry to buy technology simply because it was new or in vogue. Today, technology must provide business value to the more technology-savvy consumer.

Secondly, because today's society has a greater awareness of technology, the CTO is in higher demand from the trade press, analysts, and customers. Up to half the CTO's time may be spent on customer briefings, customer visits, and other external activities.

Finally, the CTO is no longer considered a single entity making decisions in isolation while sitting alone in a cubicle. In today's complex world, the CTO must be a capable, well-spoken businessperson who can adequately discuss relevant technology issues with both professionals and laymen in a wide variety of different contexts.

Tom Bishop is vice president and chief technology officer at BMC Software. A recognized and award-winning chief technology officer, Mr. Bishop is responsible for product vision and direction at BMC, including advancing Atrium, the company's innovative open-architected foundation for its business service management solutions, while continuing to drive breakthrough innovations. Most recently, Mr. Bishop served as chief technology officer of VIEO Inc., where he was named "Chief Technology Officer of the Year" by InfoWorld *magazine, and he is the former chief technology officer of IBM Tivoli.*

Mr. Bishop has more than twenty-five years of industry experience and has served in a variety of senior technology and strategy roles throughout his career. He is well known as a technology innovator, holding nine patents in fault-tolerant computing, and leading the development of industry standards such as the Distributed Management Task Force

(DMTF) and POSIX. As chief technology officer of VIEO Inc., Mr. Bishop pioneered the architecture and design of an applications-focused, quality of service-oriented enterprise management solution. During his tenure as chief technology officer at IBM Tivoli, he created and led Tivoli's distributed systems management technology and products.

Mr. Bishop also serves on the board of directors of Witness Systems. He holds bachelor's of science and master's of science degrees from Cornell University.

Delivering the "Wow" Factor

Ian M. Marlow

President and Chief Information Officer

Global Facility Services

A CIO's Vision

The personal vision of the chief information officer (CIO) should be to bridge the gap between the business world and the technology world and provide the ultimate balance between the two. Success is based upon having a strong network, a talented network team, and cooperation from everybody in the company. There are several qualities a CIO needs to be successful. First and foremost, a CIO needs to be a people person. He or she needs to be technology- and business-savvy. The CIO needs to be approachable and have a sense of humor. He or she also needs a financial background in order to quantify the return on investment on any given project and properly explain those returns to the management and the board of directors.

The CIO and Public Relations

The CIO should have strong public relations abilities. The CIO must be able to understand how to market projects and show the return to skeptics within the company. The skeptics need to be awestruck by the projects, and the best way to achieve that is to show them how the project will be accomplished, let them play with the new technology, and provide them with an example of something the CIO has accomplished for another company that had an impact on the business. It would also help if the CIO could impress upon the skeptics how happy the investors were for the previous company by offering examples of praise the CIO received.

The Challenges of the CIO

The most challenging aspect of being a CIO is people not understanding the value of many of the things the CIO is trying to do. In many companies, the CIO and the chief financial officer (CFO) butt heads because the CFO wants to see financial gain every year on a new project when the CIO may recognize that it will take longer than twelve months for the project to produce returns. Often, a CIO will find that a project has been eliminated because the CFO is overly worried about the current cost of the project and does not care that the long-term return on investment far outweighs the current cash outlay. However, because so many implementations fail, the CFO may have a healthy skepticism for new projects that is entirely justified. Most projects fail due to changes in personnel and/or leadership, changes in consulting groups, incomplete documentation, and poor project planning. The challenge for the CIO is finding the right way to overcome the skepticism and earn the funding to get the project to its conclusion while keeping the team intact to deliver the project in functional form and train the end users for whom the product(s) are designed.

Being Prepared

The CIO must always be prepared to help the company make technology decisions. If the company does not have a steering committee, then the CIO must be prepared to play the role of both the chief operating officer and the CIO within the company. He or she must have an in-depth understanding of the business and understand what it takes to run the business. Because the CIO understands the company needs from a financial aspect and from a user standpoint, he or she is best able to pick technologies that work well and are cost-effective. The CIO needs to understand where the market currently is for the company's industry. For example, if the company is in the real estate market, the CIO might be part of the building acquisition and/or disposition phases of the business deals. In the disposition phase, the CIO needs to worry about gathering and providing accurate historical and financial building- and investment-related information. In the acquisition phase, the CIO needs to worry about installing and supporting technologies to help ensure that the underwriting process is sound. The needs of a company are somewhat cyclical depending upon the market, so the CIO must constantly be prepared with the information necessary to make the best decisions for the company.

Selling the Technology

Once the decision to use a technology has been made, the CIO must always sell what he or she is doing in order to gain corporate support. Supposedly, Jeff Hawkins created the initial model of the PalmPilot out of a block of wood. He would sit on a park bench and simulate using the technology on the wood block to see if people would be willing to use such a tool. That is the way a CIO should look at potential technology. If other people don't believe in the project, then even if the installation is a success, the project will ultimately fail because the technology won't be used. This refers specifically to the end users and their management. The CIO must always prove the concept, either in the concept stage or the demonstration stage, and try to keep the time of the project to fewer than six months. The six month rule is a general guideline that is long enough to install technology that will impact the business model but short enough to maintain both the corporate emotional and financial backing necessary to deliver.

Direct Financial Impact

When making technology decisions, there are three specific things a CIO can do to have a direct financial impact on a company. The first thing the CIO can do is to automate process. A process should be evaluated, and then the CIO must determine how many pieces of technology might be required to automate that process. Automation is key if you can increase an employee's work time by giving them back X hours per week by reducing the tedious portion of their jobs through automating repetitive tasks. The second thing the CIO can do to have a financial impact is to continue to move away from telephone usage to Web or wireless devices, thus reducing staffing needs and speeding up reply time. The third thing is to really advocate the "work from anywhere" scenario. For example, employees might be given Blackberry devices in order to wirelessly access e-mail at all times, while others get online accounts allowing network, e-mail, and telephone access from any computer in the world.

Technology Decision Strategies

The main strategy that leads to success in making technology decisions is to constantly talk to people about their needs in order to determine what technology is working and what is not. Another strategy is to always be willing to try a new toy or technology. Only one out of every five or more new technologies will be right for a company, so the CIO must have the budget necessary to look at and test new technology before deciding if it will work for the company. The CIO must be willing to be bold and spend a little bit of money to do things, and then be able to show a return on investment. In many cases, the return on investment for new technology might simply be increasing an employee's ability to work 20 percent more because they're working wirelessly or more efficiently.

Three Rules for Success

There are three rules a CIO can follow to help him or her successfully make technology decisions that the rest of the company will agree with. The first rule is that the CIO must be able to explain the technology in terms that non-technical people can understand. The second rule is that the CIO must be able to explain the return on investment. The third rule is to always have some "wow" factor.

Due Diligence

An information technology (IT) team should always perform due diligence on potential technology expenditures. The team should first create a basic internal spec of what it is looking for. Then, the team should perform a request for proposal. Next, the team should give the spec to three or four possible vendors and let them come back with a full proposal. For example, a company Web site might have a great content engine but may not look appealing. The IT team would create a business spec of what the company would like the Web site to look and feel like, and put that information out to a number of companies for a bid. The team should interview the prospective companies, look at examples of work they've done for other clients, and try to get a fixed price before making a final decision.

Difficult Decisions

It is a misconception that CIOs view technology as a toy to be played with. The IT team is usually pictured as a bunch of geeks sitting in a back office playing video games. However, those misconceptions hide the truth, which is that CIOs take technology very seriously. The most difficult types of decisions a CIO can make regarding technology are the ones that stem from a good idea and a great deal of interest but lack the funds to start work immediately. Employees put their heart and soul into new ideas and new technology, and it is very difficult to explain to someone that their project might not be a priority for the company. It is important not to let the employee get depressed. The CIO should try to explain the reasoning behind the decisions, list the priorities that come before the pet project, clarify why they are more important for the company, and explain how the employee can bring value to the company in other ways.

Recent Changes in the CIO's Role

The role of the CIO has changed over the past few years as the job has become far more centered on business decisions. A few years ago, a person could have the dual title of CIO and chief technology officer (CTO). However, as organizations grew, they realized that they needed a businessperson and a technology person. The information officer now provides the information, and the technology officer provides the technology. As technology becomes more driven by business, many CIOs have started to physically run the organization, understand the business model, and know how technology applies to the revenue-generating area. In the future, more CIOs will be promoted as chief operating officers and ultimately as chief executive officers.

Three Golden Rules

There are three golden rules for CIOs in terms of making decisions about technology. The first rule is to generally never embark on a project that will take longer than six months to implement. The second rule is to gain corporate sponsorship before starting a project. The third rule is to always find something with which to wow people. Impressing people will make them envious of the technology and will also show the company that the CIO is keeping an eye on new technology.

As president and chief information officer, Ian Marlow's responsibilities include the oversight of a global operation spanning thirty-two countries and forty-eight states. He has direct oversight of Global Facility Services and its day-to-day operations, which administers over 54 million square feet of prime commercial and retail space around the world.

Mr. Marlow centralized Gale's operations throughout the world into a state-of-the-art, 46,000-square-foot worldwide headquarters; opened new offices in New York, New Jersey, Massachusetts, London, and Seoul; built a new data center incorporating the latest technologies, including an Internet-based network accessible from anywhere in the world; and began the implementation of the technologies that will lead the company to a paperless environment. At the same time, he reduced the company's operating expenses significantly.

Mr. Marlow's operational experience spans multiple industries and disciplines. With formal training as a chemical engineer, he began his career with the Department of Defense and Department of Energy as a nuclear engineer. In 1997, he entered the private sector, ultimately working as the chief engineering and marketing officer of one of the nation's largest publicly traded insurance groups. In April of 2002, he came to Global Facility Services and is now responsible for Gale's global operations.

Mr. Marlow earned a bachelor's of science degree in chemical engineering from Rensselaer Polytechnic Institute.

Dedication: *I dedicate this chapter to my parents, David and Ann. It is the opportunities they gave me early on, through conversations, guidance, and business acumen, which have allowed me to be confident that I had the tools necessary to lead, learn, and grow. This chapter outlines many of these ideas, which have led to helping organizations and people maximize their days and efficiency.*

The Technology Executive: No Longer a Back-Room Support Function

Wayne I. Thompson

Vice President, Information Services and Technologies

University of Medicine and Dentistry of New Jersey

Main Focus

My personal goal as a technology executive is to enable the enterprise. The objectives of my position are to add value to the organization both strategically and tactically. That means getting the right information at the right place at the right time and achieving efficiency in doing that.

Making Decisions

The decisions I help my company make involve strategic decisions, directions, foundation, infrastructure, and applications. From a team's perspective, there's a range of activities that include the tactical to the strategic. But from my perspective, it's primarily the strategic decisions or decisions with strategic implications that I help the company with.

Increasing efficiency, decreasing costs, and enabling revenue enhancement are three major aspects I am involved in that directly affect my company financially. Increasing efficiency usually goes right to the bottom line, while process improvements, automation, and all that technology brings to the table help improve the efficiency and tracking of any business process within the organization. As far as decreasing cost is concerned, if there are instances where we can use automation, technology, or improved applications to make a particular function, process, or service less costly, we try to do that. For example, health care operations can become more efficient and thereby reduce costs. Enabling revenue is probably the most elusive function, in that we perform a whole host of strategic activities that position the organization to generate more revenue. One example might include distance learning infrastructure at a university that would allow us to capture more students and educate them without expansion of physical space. We may enhance revenue generation capability by automating aspects of radiology with a PACS system that would allow us to read film and cases from other parts of the country or the world, enabling revenue enhancement in that regard.

What It Takes to Be Successful

Being a quick study on any given issue, thinking strategically on behalf of the organization, and being able to make decisions in instances where you don't necessarily have 100 percent of the data available to you are of critical importance in order to be a successful technology executive.

There are no real specific methodologies or strategies that guarantee success when making these types of decisions. Each scenario calls for different strategies and methodologies—from coalition or consensus building, to more directive-driven decision making. A combination of these skills and abilities in your management toolbox allow you to do well in such an environment. Identifying the value equation for any given decision is critical to my job, and so is helping keep focus on what that value decision is throughout the process.

Most challenging is having to balance what are sometimes diametrically opposed needs with regard to our organization's multiple missions. Sometimes, those missions require competition between the needs of internal units, which makes balancing those missions one of the more challenging aspects of the job. Other challenges have to do with the general realm of change management. These are challenges we're used to dealing with as technology leaders—from system design and planning to introduction to shepherding it through successfully to its evaluation after the fact—but that doesn't make the challenges any easier.

Team Effort

An example of a teamwork challenge we have faced was the need to fund some major infrastructure improvements in an environment in which the funding (budget) was distributed to units. In order to move forward, we needed to convince eight units and six other business areas that this was an idea worthy of their budget dollars. This required a series of meetings in which we first educated the group about the challenge and then invited them to participate in developing potential solutions. Everyone's participation in finding the best solution engendered a sense of ownership and enabled us to make progress.

I have developed a governance structure that is comprised of committees for each mission area. In my case, that means education, research, health care, and administrative. These committees help to formulate strategy and policy. I also work closely with the chief executive officers of our health care units and the deans of the eight schools with respect to strategic decisions regarding technology that need to be made. Also included in the decision making are the chief financial officer of the organization and vice president for facilities and planning, who are usually involved in major projects such as new buildings, renovations, or upgrades to the environment.

With respect to understanding domain issues, there is a greater expectation for me to understand the business of my fellow deans and vice presidents than it is the other way around. They expect me to understand their realm and some of the challenges their businesses face in order for us to be successful and present them with viable solutions. That's something I focus on—making sure I understand the main points and challenges of those business units before discussing a solution with them. On the other hand, I think there is less of an expectation on my part that they have a deep understanding of the technology realm. I do have an expectation that there's a general understanding of some of the challenges of a services and support organization (e.g., purchasing, facilities), but not necessarily a deep understanding of the technical environment here.

Goals are formulated and reviewed by a system of checks and balances. This is achieved by tying together our annual goals with the strategic plan and objectives of the institution as one source. A second source is our governance structure: We have four committees that outline priorities from each of our mission areas. Thirdly, we have internal or information technology (IT) organization-specific goals such as infrastructure, which probably would not come directly from a mission approach or strategic planning. Those three sources form the basis of our goals. We monitor activities and specific projects within the year relating to those goals. Our ongoing monitoring of those projects and quarterly review ensure that we are making the expected progress.

Good Communication is Necessary

One piece of advice I often give is to over-communicate regardless of the circumstance. Good communication often sees us through many a storm and certainly makes for successful projects. Over-communication is at worst an annoyance, but under-communication usually guarantees certain disaster. In terms of technology decision making, the advice I give to staff is the same: You can remove technology from the equation and the communication advice stands on its own regardless of what you're talking about. For instance, if we are pursuing a particular solution, it's still necessary to communicate with everyone from the affected constituency, to sponsors and executive sponsors. A better piece of advice I have often used is that culture eats strategy for lunch. Particularly in the environment of the academic medical center, it's wise to keep in mind at all times that what is possible is not always what is tolerable.

Strategies and Goals

Establishing a governance structure for the IT organization is absolutely critical. Being able to build constituency or mission leadership alignment and participation makes the job inordinately easier. When it comes down to making decisions, you're making them in concert with the organization as a whole, which aids in ensuring that the IT organization is aligned with the mission-based and business unit needs. The technology executive has a built-in, full-time job, which includes educating fellow executives and peers, leading the organization with respect to the value technology can bring to the table, and taking on the challenges of maintaining and supporting a technologically intensive environment.

One of the most difficult technology decision-making challenges is a constituent who already has a strong bias or preconceived notion of which solution is best for the environment. They come to the table with that notion, making it difficult to convince them otherwise if they have the budget to move forward. Your job is often to help them see the wisdom of going through a due diligence process or to become open-minded to the possibility of some other solution. The other difficult type of decision is one where there are several solutions that have much of the base needs covered but only have promises to cover the rest of the needs in the near term or the infamous next release. In such cases, you are taking a gamble that the vendor will deliver on the rest of the product.

Technology Decisions and Return on Investment

Calculating return on investment with respect to technology decisions includes making sure we have a good understanding of the costs of whatever the solution is. In our environment, we typically separate one-time costs such as implementation or acquisition of hardware from the recurring costs such as people, maintenance, technology, and so on. Once we have a good handle on the costs, we move on to the revenue side in terms of what we think it's going to generate, which usually contains both hard and soft components. The hard components are usually easy to quantify. We usually take a conservative approach to whatever we believe the soft side is. Or in some cases, we weight the soft side lower than the hard side. Those are applied against the cost and financing methodology to determine whether there is a payback or return on investment.

Evaluating Previously Purchased Technologies

There is an annual evaluation of any new systems that have been implemented, but only for the largest of our investments. We do not, however, have a formal routine review of medium or smaller investments we make. On the front end, for a proposed technology investment to be worth considering, a contribution to the strategic capabilities of the organization is necessary in addition to delivering strong value in some way and fitting into our standards and technology profile as an organization.

One of the biggest misconceptions about making technology decisions is that it is simply making easy A versus B comparisons or that you can have a "shootout" among two or three solutions and come up with the best one. This misconception persists quite possibly because of the popularity of PC magazines and other media that contain comparison charts with scores and two solutions side by side. Key decisions simply do not fit into that model when you're talking about strategic investments. The misperception also persists because it is easier for people outside of technology to get a general understanding of it (although not very in-depth). Everyone understands an A versus B comparison with a few checkbox criteria. In regard to my position as a technology executive, there's a misconception that this realm is far less people-intensive than it actually is. At the root of all of our solutions are hard-working professionals who make them work.

On Being a Better Technology Executive

In the past few years, the role of the technology executive has changed in that it has become much more strategic in larger organizations as opposed to a back-room support function. It has become more vital to the operations and well-being of most organizations instead of being a consideration after business transactions have occurred. What I have found most useful in making technology decisions in my position is membership in professional organizations, many of which have a bulletin board type of service and periodic conferences. Such organizations often do research and condense important information, which brings it to you quickly and precisely and helps you do your job.

My advice for technology executives when making technology decisions would be that they should make sure they can engage stakeholders. Make sure you've defined the question well. Set expectations. Define success. The phrase "Culture eats strategy for lunch" is particularly applicable to large, decentralized organizations like academic medical centers, but it also applies to almost all other settings to some degree. Be sure the cash flow through the entire lifecycle is understood and agreed to before you advocate a solution. Senior leadership should understand the solution and its value so they can be supportive during the inevitable bump in the road. Lastly, make sure you have contractual weight behind any solution. It is an option you hope never to use, but it can be a true deterrent to future problems.

As of June 7, 2004, Wayne Thompson became the vice president for information services and technology at the University of Medicine and Dentistry of New Jersey. His career in information services spans twenty years.

Previously, at Thomas Jefferson University in Philadelphia, where he served as vice president and chief information officer since 2001, he managed the information system's group for the Health Sciences University, which has three colleges and a faculty practice group.

Prior to joining Thomas Jefferson, he served as chief information officer at the University of South Florida Health Sciences Center in Tampa, Florida. He also serves on several professional committees, including as chairman of the group on information resources benchmark committee of the Association of American Medical Colleges and as a member of the Philadelphia board of directors of the Society for Information Management.

Generating Positive Results for the Business

Scott A. Storrer

Executive Vice President,

Service and Information Technology

CIGNA

My Role and Responsibilities

My position is a joint role leading both service and technology for CIGNA. CIGNA is shifting its business model to be a leader in consumerism, health advocacy, and clinical informatics. That business model change and the accompanying strategic focus, service, and technology are critical to success.

When I stepped into the chief information officer (CIO) role, we were coming out of an era where our information technology (IT) organization was not as well connected with the business as we need to be as we move forward to this new model. My organization has been working very hard over the last twelve months to gain better alignment and forge a stronger partnership with the business. Three years out, I want IT to be one with the business.

The role of the CIO in the IT organization is to bring new business capabilities to life. My responsibilities are to understand the needs of the business and to work closely with the business leaders to crystallize strategy into the operating plan so we can achieve our business objectives.

As the business evolves to a consumerism strategy, we need to think about our business capabilities and the business applications that are needed. In order to power those business capabilities, we need to upgrade and strengthen our business and IT architectures and infrastructure—to make sure we have the framework, wiring, and plumbing in place to support future capabilities. My greatest job challenge—and opportunity—is making sure our organization is aligned with the business in this regard.

My Decisions

I help the company make technology decisions. From a technology standpoint, we have to improve our architecture, which is the framework that supports all of our business applications. We also need to improve our business applications so we are member-focused and user-focused. Our infrastructure, architecture, and business applications need to support consumerism. There is a lot of heavy lifting underway in our organization to upgrade our architecture and business applications.

We are adapting and enhancing our architecture to support consumerism, health advocacy, and informatics. Essentially, that means we are changing our data structure and being more member-oriented. A service-orientated architecture allows business applications to post data from the same sources. This helps share data at the member level and allows us to essentially arrange it to be shared across multiple applications. We can make sure we have absolute consistency with members, employers, and providers, which goes a long way toward fueling self-service capabilities.

When we got into the various applications that enable consumerism, health advocacy, and informatics, we found that we needed to develop more robust decision support tools for members to make health care purchase decisions. We are working aggressively to strengthen our existing Web-based capabilities.

Consumerism

The trend is for consumers to pay for a greater share of their health care. As a result, we want to bring absolute transparency to our members around the services they would purchase in the marketplace. If a member is having knee surgery, we want him or her to be able to go to a CIGNA Web site and identify five different doctors who can perform knee surgery along with their total costs. At the same time, we want to provide the quality scores that are associated with work the doctors did in the past.

When it comes to health advocacy, CIGNA believes in holding hands with our members and helping them navigate the health care environment. The health care industry is fragmented and has many stakeholders: There are primary care physicians, specialists, hospitals, and other facilities. As we move through the world of consumerism and as consumers purchase more and more health care out of their own pockets, not only is transparency important, it is also important to have a nurse case manager to guide people through the health care process.

Health Advocacy

In the area of health advocacy, we are investing in the business capabilities that will essentially enable our nurses to provide advisory services to members. A number of our largest customers have purchased CIGNA health advisor services for a large population of their employees, and we have built a Web-based portal at the individual level.

An employee with a health issue can call a health advisor who can help channel the employee to the best specialist from a price and quality standpoint. The health advisor maintains communication with the member throughout the process. CIGNA has 1.5 million members in a health advisor program, and we believe that over the next three years the majority of Americans will have a health advisor service.

Challenges

IT is a young industry. We are only forty years old. When you look at management programs like M.B.A. curriculum, only in the last few years has IT gotten an active voice as part of training and development. The degree of understanding of IT from business owners, sales, product development, and service employees is mixed. Business leadership needs a strong command on how to best leverage IT assets and manage IT projects.

In the new world, our vision is to drive technology. Our projects are not owned by IT. They are owned by business leaders. The greatest challenge I have is that on the business side we don't have business leaders that are as savvy as they need to be with IT accoutrements. That will change over time as more CIOs come from business rather than technology. I didn't grow up in technology. My whole career has been running companies with a profit and loss or service focus. We will see more CIOs coming out of the business, which means tighter connections with the business. Over time, business leaders will have a greater degree of ability when it comes to managing the IT connection.

Another challenge lies in the strategy we have around consumerism. The products and services in the market today are at a beginning stage. If you think about the advent of HMO products, which ran fifteen years, the time was right for the market then, but market need is evolving away from the payor being the gatekeeper for care to consumers having more choice and responsibility.

We are at the very beginning of our product capability in the marketplace, and the impact it is going to have on reducing the overall costs of health care is unknown. From that standpoint, we are making large bets on technology we deploy to fuel business applications. CIOs of manufacturing companies tend to focus on inventory management systems technology. When it comes to powering consumer-directed tools, there are no existing technologies. There is a lot of work dedicated to making conscious decisions about what we can build on our own. It comes down to whether we have the IT knowledge within our organization to accomplish such tasks.

At the same time, we have to keep a savvy eye on small startups with certain capabilities that could empower our solutions or business applications. One of the gaps we had earlier this year was that we did not have the ability for members in consumer-directed products to model their health plan. Essentially, these members moved away from another plan to a consumer-directed plan, and based on what they signed up for they could have dramatic swings in what they pay to support health care. We had a gap as far as our ability to provide decision support tools in the marketplace, and so we acquired a company named Choicelinx whose area of expertise complemented our needs.

Setting Goals

My number-one goal for the team is to generate results for the business. We generally do three major releases a year where we bring new technology into the environment. With our old model, we defined success as delivering all of the capabilities we said we would. Now as we move forward, our emphasis is more focused on measuring the results we generated for the business.

Our second goal is to make sure our IT strategies are fully aligned with the business strategies. Our third goal is to make sure we are building the capabilities we plan. Our fourth goal is to make sure we work within the investment parameters the business can afford. Anything we do has to tie in to the strategic financials of the business.

We have an intensive governance process that is business-driven. We have a governance council that is led by the president of the company. We also have an executive management team that checks the status of all of our projects on a monthly basis. Any project worth more than $2.5 million or identified as a large business benefit will be reviewed by the team.

We also have governance around our development process. There are certain rules we follow to make sure our methodologies are consistent across our organization and also with leading industry best practice standards. Underneath that, we have architectural governance to make sure we are not introducing harm into our environment and that everything we build is going to support our applications as we need them today and in the future.

Advice

I approach all of my meetings not as a technologist but as an equal business partner. The number-one piece of advice from our chief executive officer has been to realize that even though I know the business, the business is not as savvy on technology issues as it should be. It comes back to putting discussions into terms the business will easily understand and using those discussions as opportunities to educate the business leader about the technologies we are deploying so they see how critical IT is to the future of the organization.

Biggest Expenses

The biggest technology expense my team incurs is improving our architecture through data and service-oriented programs. Our second largest bucket of expenses is capabilities development and business applications. The third largest bucket is our data center operation. Our business information protection is an area in which we see funding double year after year because of the virus threat, privacy, and offshore activities that open an environment to other vendors.

Due Diligence

My team does due diligence on potential technology expenditures the same way any potential acquisition due diligence occurs. We first make sure we have a business leader. One example of this is our relationship with Choicelinx.

When we purchased Choicelinx, the product manager was responsible for consumer-directed projects and led the charge on analysis. When it comes to IT, I make sure the team has adequate representation between applications development, architecture, and data center operations so that when they look at a piece of technology they can determine whether it will fit into our overall architecture and whether it is going to create harm.

From a development standpoint, we determine whether we have the skills in-house to develop something and whether we can leverage it within our offshore model. Finally, we drill all the way down to the data center to make sure the technology will run efficiently throughout the company.

Evaluating Previous Purchases

We evaluate previously purchased technologies frequently. In 2002, we implemented a new technology platform for our entire business. Three years later, I continue to update our senior management team on how these new end-state systems are performing. Because this IT project completely transformed our business, it's a dramatic example. Nevertheless, many of our projects in the areas of consumerism, health advocacy, and informatics will be reviewed twice a year for the next three years.

Misconceptions

One of the biggest misconceptions about making technology decisions in my position is that technology is easier than it actually is. A lot of the misconception is due to retail technology. When people purchase an iPod, they are ready to use it within an hour. The technology used in the business community is much more complicated.

In our business, people think there are technologies out there ready to support the marketplace of consumerism, health advocacy, and informatics. There are not. There are emerging technologies, but I can't find one open-ended solution to support all projects. I have to put together a lot of small projects and small solutions. That is why architecture is so important.

Another misconception is that offshore capability is a bad thing. We are steadily increasing our application maintenance and development work at near-shore and offshore locations. This allows us to work around the clock. Offshoring is not just a cost play; it is a quantity play. We can bring more to market through faster releases, and we also have the ability to acquire skills that are rare in the United States.

Golden Rules

I just entered my role as CIO last November, and my number-one rule for success has been to get involved with the business strategy. My number-two rule is to educate my business partners on how to leverage the strategic assets of IT and how to manage projects. My third rule is that once I begin a project, I make sure it is business-led. Everything we do should be through the eyes of the business so that what we deliver at the end of the day is what the business requires. I never underestimate the leverage that service-orientated architecture can bring my business.

Scott A. Storrer is executive vice president of service operations and technology for CIGNA Corporation. In this role, he is responsible for consumer, employer, and provider services, process management, and strategic initiatives for CIGNA, and for the corporation's systems development, data processing, and telecommunications worldwide. Mr. Storrer was named to this position in June of 2005. Mr. Storrer joined CIGNA in May of 2001 and became senior vice president of disability management solutions and customer service for CIGNA's group insurance division. In October of 2002, he was named senior vice president of service operations for CIGNA HealthCare and has served as interim head of CIGNA's information technology organization since November of 2004.

Previously, Mr. Storrer was with Liberty Mutual Group, where he held a variety of senior leadership roles of increasing responsibility during an eight-year period. Prior to joining Liberty, he was with General Electric Company, where he was a graduate of the General Electric Financial Management Program.

A graduate of DePauw University, Mr. Storrer earned a master's of business administration degree from Boston University. In addition, he has attended several executive and financial management programs in the United States, Canada, and Japan.

Going Beyond the Traditional Limits of Information Technology

Bill Chapman

Senior Vice President and Chief Technology Officer

Avnet Inc.

Personal Vision

The vision Bill Chapman holds for his organization reaches beyond the traditional limits of information technology (IT). He sees the chief technology officer (CTO) as a proactive partner in the development of strategies and agile solutions that accelerate revenue, growth, and market differentiation while increasing profitability. Most of his decisions revolve around the applicability of information and business technologies, and how associated tools and processes can be leveraged to create differentiating business value for the various Avnet business units.

A broad vision for the office of the CTO is in place. "I have to ensure that the spectrums of technologies needed to support the businesses are in place. As we look to the future, every step along the way has to be compatible with the working environment. The transitional phases and associated deployments of the overall architecture are incredibly important. Our infrastructure and application teams provide significant value for Avnet in the marketplace, and at the same time maintain a viable and evolving business environment."

Creating Financial Impact

Avnet is a technology distribution company with business relationships reaching throughout the electronics industry supply chain. It distributes and adds value to the products of more than 300 of the world's leading suppliers of electronic components and computer products. Those suppliers are considered to be "customers" in terms of how much importance the company places on those relationships. The actual customers who purchase from Avnet are the world's original equipment manufacturers, contract manufacturers, value-added resellers, and system end users. Customer- and supplier-facing applications have enabled the company to bring new value into the marketplace and have aided its growth.

When Mr. Chapman first came to Avnet over five years ago, the company had undertaken a major initiative to install a new enterprise resource planning (ERP) environment and to exploit the Web. Having just experienced the Y2K frenzy, Mr. Chapman suggested the company delay the replacement of its ERP environment and instead focus on the Web and customer self-service.

As a result, the company deployed an isolation layer similar in concept to enterprise information optimization. Avnet's isolation layer utilized techniques from both middleware and operational data stores to isolate its ERP systems from front office applications that face employees, customers, and suppliers. "We focused our IT organization on creating differentiating value for our businesses rather than turning our business environment upside down by deploying a new ERP environment during an economic down cycle," says Mr. Chapman. "As a result, we saved millions of dollars in IT expenditures, created new revenue streams, and gained market share by focusing on the marketplace with an emphasis on new value to suppliers and customers during a down market." Indeed, in North America, Avnet's computer marketing business grew from approximately $1 billion to $3 billion in the last five years.

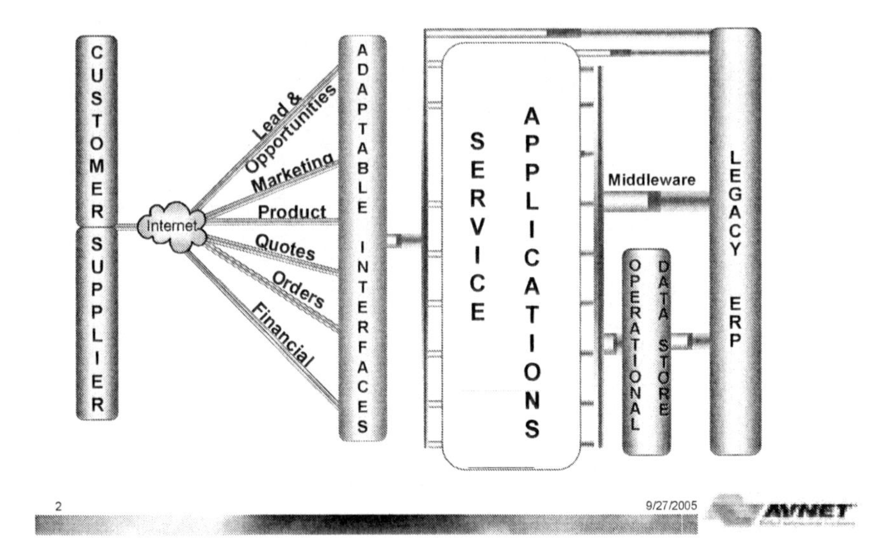

Avnet was the first company to do true large-volume business transactions utilizing RosettaNet to enhance order processing and purchase ordering. "We did that with one of our largest suppliers, HP, in 2000," Mr. Chapman says. Avnet developed a supplier business-to-business capability based on this same isolation layer concept used to do customer-facing applications. Using a common architectural approach for suppliers and customer-facing applications enabled Avnet to provide new value to Avnet's partners and thus gain significant market share.

Biggest Expenses

The biggest expense for the majority of IT organizations is the investment in the infrastructure, even if a company outsources or has its infrastructure hosted. When Avnet assesses new hardware, its infrastructure support team makes sure the new product makes sense for the types of applications that will be running on it. Some applications run well in a distributed environment where there is a need for network bandwidth and significant requirements for memory. As a CTO, creating a synergistic relationship among the infrastructure, architecture, business process, and applications teams is critical. The relationship should include architecture reviews, a change control process, clear communication procedures, and of course, a congenial work environment.

The Artistic Side

According to Mr. Chapman, "The art of being a CTO transcends knowing everything technical. It's about knowing how to apply technologies, design solutions, and address market needs using technologies and business processes to add value."

To be successful, Mr. Chapman believes a CTO needs several personality profiles in order to maintain congenial working relationships with many different people. For example, the CTO works with technical people who have certain personality traits and with business leaders who have very different personality traits. The CTO must be able to adjust communication styles on short notice.

The CTO is a middle person in the organization who needs to understand how to use technology but doesn't actually have to know every nuance of every technology. The job is all about matching the right person and the right technology to the opportunity.

A vision is essential. Some people want to talk in abstractions, and other people need detail and structure. Much of the CTO's success depends on understanding how to use technology, how to design solutions, and how to successfully convey thoughts to different constituents.

Strategies for Success

In his position, Mr. Chapman believes every individual is critically important. "We hire people because they add value to the corporation. People on the phone taking orders can offer great insight. Senior executives provide our overall vision, but the IT team gathers information from all levels of the organization. A great company makes sure everyone feels like they're part of the team. Many, if not most, of the opportunities are foreseen by individual contributors."

From a return on investment (ROI) standpoint, it's easy to obtain twelve-month ROI cycles on new technologies that are adjunct to core foundational systems. As Avnet addresses foundation pieces and legacy renewals, payback cycles slow down. It's a matter of business opportunities and a partner perspective. According to Mr. Chapman, most of the significant paybacks are achieved with projects that supplement or combine fundamental architectural pieces. So, from a CTO perspective, the company should not only visualize the ROI of a foundational piece, but the potential value of the long-term supplemental pieces, all viewed from a business perspective. "If a CTO or IT organization fails to view itself as a valuable partner of the business, there's a problem. At Avnet, we look at IT as a fully qualified partner with the businesses. We have the same responsibility as any other business unit or individual within the company."

Challenges

The challenge of a CTO, as Mr. Chapman sees it, is to remain nimble. The CTO must select an approach or combination of approaches to address each individual problem. "We have actually trained senior members of the IT department on different design and problem-solving techniques." Once that approach is selected, the challenge is to look at a composite perspective of the future environment with transition in mind. Next, it's necessary to drive down into particulars on how the actual application is designed to run. What are the features? Where are the opportunities that enable the company to differentiate itself in the marketplace?

Mr. Chapman thinks about differentiation as if he were a car designer. "If I were going to design a car for 2012, what would people want? Some people might say it's all about the real estate on the steering wheel. They want audio controls, cruise controls, and everything accessible on the steering wheel. Someone else may not care about the steering wheel. That person wants voice recognition. All the features on the steering wheel available today may be available using voice recognition in the future. To me, that's what has to happen in an organization—you need to get dynamic people together and generate ideas openly. That collaboration will create a change in perspective and identify opportunities to create paradigm shifts in the marketplace."

The most challenging aspect of being a CTO is working in collaboration with everyone else. It's easy for people to think of the CTO as a consultant to the organization and drop him or her into the "policing arm" bucket of an organization. The key to success is being much more open-minded and team-oriented. "It's about being hands-on, achieving commitments, and adopting them in the day-to-day culture."

The Team

When making technology decisions, Mr. Chapman goes to technical leaders and functional teams to obtain abstractions. "Many solutions in today's businesses are combinations of fundamental technologies and incremental solutions. There are very few fundamental technologies that would create breakthrough market impacts, but combinations of various fundamental technologies and incremental features can readily create differentiating value to the marketplace." The CTO needs a tight relationship with the business leaders of an organization so they are in touch with the business's holistic needs. When the CTO puts a solution in place, it has to be viable and applicable to real-life situations.

The best piece of advice Mr. Chapman could provide his team concerns how to approach design. There are quite a few different models for approaching problem solving. Understanding different approaches and applying them, based on the person or opportunity being dealt with, is essential. Solutions must be described from the perspective of each audience member. Additionally, different problems lend themselves to different design approaches. For example, some individuals readily understand stepwise refinement of complex ideas, where other individuals relate to object abstractions, pictures, outlining, forward-based designs (opportunistic), or backward-based designs (goal-based).

Additionally, checks and balances in a team environment ensure that the team's organizational goals are met. Avnet substantiates an ROI perspective on initiatives. It's not all about research and development or doing things at the lowest cost. ROIs are about generating profit, satisfying regulatory requirements, creating market barriers, and extending the life of fundamental business solutions (i.e., maximizing ROI and creating the biggest impact on customers, suppliers, employees, and investors). "We have a robust process where we identify opportunities and do basic analysis and investigation. We develop a project initiation document with ROI studies exposed. Based on that information, we govern our plans to move forward."

As senior vice president and chief technology officer for global distribution giant Avnet Inc., Bill Chapman directs strategic information technology initiatives across Avnet's operations worldwide.

Mr. Chapman joined the company in November of 1999 as vice president of information technology, became senior vice president in August of 2001, and was named group information officer in March of 2002. He led the successful deployment of advanced system architectures, which enabled the rapid and effective assimilation of companies Avnet was acquiring, and which were critical to its growth strategy.

Prior to joining Avnet, Mr. Chapman worked for Motorola for more than twenty years. As senior director of information technology, Mr. Chapman had responsibility for more than 400 information technology professionals supporting computer-integrated manufacturing, supply chain systems, and demand chain systems, as well as SAP and architecture. Earlier, Mr. Chapman was director of Motorola's computer engineering development group, where he was awarded eight international patents. Mr. Chapman's Motorola career began in the Semiconductor Research Lab, where he was in charge of CAD/CAM systems. He also has experience in the oil exploration industry, where he was responsible for all information technology, finance, purchasing, and warehouse operations.

Mr. Chapman received his bachelor's of science degree in management science/operations research from Colorado State University. He earned his master's of business administration from the University of Phoenix with the distinction of being named class valedictorian.

Being at the Forefront to Search for the Next Technology Wave

Arpad G. Toth

Chief Technologist

GTSI Corp.

How to be a CTO

A chief technology officer (CTO) should try to become not just a thought leader but an expert in several key interrelated technological fields. He or she should have the skills for synergistic understanding of major technological driving forces, application opportunities, and related benefits to customers and key business stakeholders. A CTO must continuously prove that he or she has a critical role in growing the business of the company. He or she must work closely with the top management including those responsible for the revenue generation, operations, and overall leadership of the company.

Making the Right Decision

The CTO must be factual about technology, trends, customer requirements, and the expected commercial value created. Specific technology decisions the CTO should make for the company include those related to base technologies and cutting-edge product initiatives. Specifically, the CTO should be a key contributor to corporate decisions to create a new vision and direction for the company as driven by technology initiatives. The CTO should have the ability to employ his or her business skills to integrate various technologies, create viable business scenarios, and present them to the executive leadership for final decision.

Making a Real Financial Impact

There are three key ways a CTO can have a direct financial impact on a company. The first way is to lead in the generation of the right roadmaps for applications, technology, products, services, and integrated solutions. The second way is to be the recognized technology leader for the company. The CEO, senior management, and employees of the company should be able to turn to the CTO for trusted opinions and advice. The third impact a CTO can have would be related to his or her ability to have the vision and foresight for competitive threats and challenges as driven by new technology trends.

Personal Qualities of a CTO

The CTO is an artist; he or she is a corporate leader different relative to a chief financial officer or chief operating officer. The CTO must have the ability to dream and think freely independent of corporate and business politics and at the same time define what value can be generated for the customers and key stakeholders. The CTO should have the freedom to work across the organization and to create virtual teams with new expertise and capabilities on an ongoing basis. The CTO must understand the culture of the organization and be able to fit into that culture. If the CTO feels the culture needs to be changed, then he or she has the responsibility and role to influence the change of culture but not to be the leader of that change. The CTO must be a recognized technology leader within the company. Industry-leading corporations should have nationally—and in many cases, internationally—recognized CTOs leading the technology-driven development efforts.

Strategies that Work

I have applied a simple strategic model a CTO can utilize to illustrate the phases of innovations for the purpose of success and profitability (see diagram that follows). In all situations, the CTO should start by understanding the benefits and drawbacks of old ways of applying and selling old technology (OWOT) (i.e., what it would take in keeping the legacy systems and products). The next component of my model includes how we can apply, develop, and sell old technology in new ways (NWOT). While this strategic phase requires the minimal investment and lowest risk, the expected return on investment is also the lowest. Typically, the developed scenario is used to extend the product or service lifecycle but is not applicable for breakthrough innovations and new business creation. The next component of my model would be the approach of defining old ways of applying and selling new technology (OWNT). In this phase, we expect to innovate in technology but apply it in the conventional way. Typical examples for this phase are the direct broadcast satellite service or cellular telephony. These solutions have been competing with old solutions already well established for public use. Lastly, the highest-risk but most profitable technology-driven tactic is the new ways of applying and selling new technology (NWNT). This requires the highest level of creativity and corporate investment. A good example could be the use of next-generation Internet for television broadcasting and on-demand services.

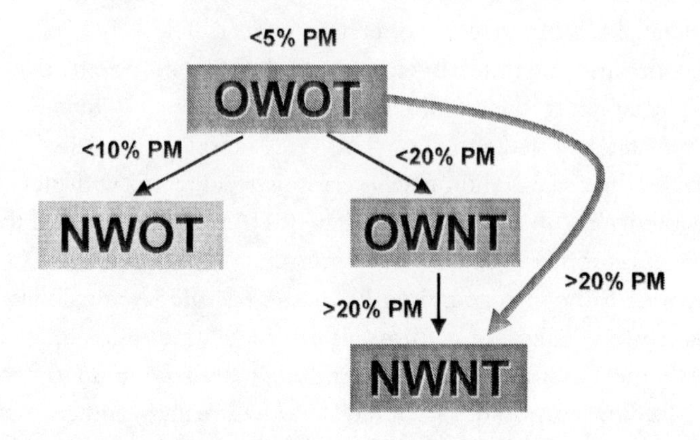

OWOT: Old Ways of Applying and Selling Old Technology
NWOT: New Ways of Applying and Selling Old Technology
OWNT: Old Ways of Applying and Selling New Technology
NWNT: New Ways of Applying and Selling New Technology
PM: Profit Margin

Challenges of a CTO

The CTO is considered a "change agent." For this reason, the most challenging aspect of being a CTO is being accepted in a corporate culture. Each corporation has its own culture. Some are on the extreme left of no change; others prefer graceful evolution and moderate changes; there are also those who are aggressive and always at the cutting edge of technology change and innovations. The best approach for a new CTO is to listen, analyze, and then get into the act. The CTO must build credibility through superior performance and build alliances with key decision makers before he or she can become highly influential for the corporation. Building credibility takes time. Throughout the corporation, the CTO should gather enthusiastic followers, those who have the skills to innovate and apply the change for the benefit of the company. One of the most challenging tasks is to find and mobilize the right resources and deliver the goals and objectives set for the future. In many cases, the CTO has no staff reporting to him or her, and must be able to cross boundaries and apply his or her skills in general management, program management, and sales and marketing management.

Be Valuable

In order to accomplish any technology goals while working for a company, a CTO must be considered an irreplaceable member of the executive team. By being valued, the CTO is able to make sure any new technology or concept developed by any employee of the company will have a business plan. Development of new technology and resultant products and solutions can be very costly. Significant mistakes can be made if the CTO draws the company in a direction that is not profitable. On the contrary, it is almost impossible to create a new technology-driven vision, direction, and business plan if the opinion of the CTO is not valued.

Changes for a CTO

The role of the CTO has changed in the past few years. The position has become far more tactical than strategic. The CTO now has to be a functional member of the leadership team; understand product and business development, sales, and marketing; and be able to present a business case. The ideal CTO must lead the company in rapidly adopting new technology and applying it to the fullest potential. This change is largely due to the fact that technology evolution has sped up and product lifecycle has shortened.

The Golden Rules

There are three golden rules every CTO should follow. The first rule is to be extremely value-driven. In other words, everything the CTO envisions and develops should serve customer needs and deliver expected profitability to the company. The second rule the CTO should follow is to be humble about results and accomplishments. The CTO is a change agent, a dreamer, an implementer, and a creator of high-impact new businesses. He or she should ideally leave the selling of the result to those who do that with professional excellence. The third golden rule is to be approachable to all employees, managers, and executives in the company.

Arpad G. Toth, a globally recognized technology and business leader, has spent his career in industries involved in cutting-edge and innovative technologies such as telecommunications, digital television, and imaging. Currently chief technologist for GTSI, he is responsible for overall information technology, homeland/physical security, and force protection technology leadership within the company. He is the founder and chairman of the InteGuard Alliance, a group of more than sixty corporations in the physical security field that provide trusted physical security protection services to governments.

Recently, Mr. Toth was elected to the ASIS International Chief Security Officers Roundtable—Government, a joint working group that consists of approximately fifty senior officials from all facets of private industry. Its mission is to give the security industry a stronger voice in business and public policy. The group is tasked with advising and guiding ASIS's active partnerships with Congress, the White House, the Department of Homeland Security, the United States Chamber of Commerce, and other major institutions that set our nation's business and public policy agendas, including those related to security.

As the head of GTSI's physical security protection technology practice, Mr. Toth led the team that provided the video surveillance system for Super Bowl XXXIX in Jacksonville, Florida, and completed the phase one design of the security protection system for the Department of Homeland Security's headquarters.

Throughout his thirty-five-year career, Mr. Toth has held a variety of executive positions with leading public and private sector technology companies. He is the co-founder of a wireless data solutions provider, served at Circuit City as chief technology officer and corporate vice president, was chief imaging architect and executive director of Eastman Kodak, and was chief scientist of high-definition television at Philips. He is a scientist, inventor, and business executive, contributing to the formulation of fundamental patents for high-definition television systems, online networked imaging solutions, and broadband telecommunications including ISDN and SONET. During his career, he has created numerous successful businesses.

Mr. Toth concluded his Ph.D. studies and holds an M.S.E.E. in telecommunications, both from the University of Toronto. He earned a master's of management degree and an electrical engineering telecommunications degree (summa cum laude) from the Technical University of Budapest. He has completed executive management programs at Harvard University and the Wharton School of Business at the University of Pennsylvania.

Aligning Technology with the Goals, Objectives, and Strategy of the Business

Richard Toole

Senior Vice President and Chief Information Officer

PharMerica

The Objective

First and foremost, the role of the chief information officer (CIO) is to look at the business needs, strategies, goals, and objectives of the company and assess and define what the company must do with the use of information technology (IT) in order to reach those goals. The CIO should make sure the goal of IT is to become world class at enabling the business to do what it needs to do efficiently and effectively through its business processes. He or she must leverage technology to help the business grow. The CIO must also offer technology at the lowest total cost, provide accurate information in a timely and responsible way so better business decisions can be made, and provide alternatives to technologies available to the business.

The Art of Being a CIO

The art of being a CIO has been evolving over the last decade. The onslaught of the World Wide Web technologies during the late 1990s and 2000, in addition to increasing business focus on competing globally, has caused the position to change from being immersed in technology and its cost to being much more strategic in that the CIO more frequently reports to the CEO and is a key member of the executive team, tasked with developing future strategies. Thus, the CIO position is now requiring more knowledge and experience in business management and less in the depth of each technology. More frequently, companies are filling the CIO role with business executives rather than pure technologists as in the past.

Qualities of a Successful CIO

There are specific qualities a CIO needs to possess in order to be successful. First and foremost, the CIO must understand a company's business model and the role IT can play in its success. Secondly, the CIO must be able to communicate both in writing and with the spoken word. The hardest aspect of communicating is taking techno-speak and turning it into business-speak, particularly as it relates to the value IT is delivering to the business. Chief executive officers (CEOs) get frustrated and develop a negative perception of IT when CIOs cannot communicate the value being delivered in business metrics rather than IT operational metrics. This translation is difficult, since the CIO frequently deals with concepts that affect many diverse business functions at once. For example, persuading the executive committee to provide investment funds for upgrading communication networks requires the CIO to use simple, easily understood analogies such as "given the company's current situation, not upgrading the communications network would be analogous to driving from a four-lane highway onto a dirt road."

Another specific quality is being able to manage a more diverse resource base both within the company itself and outside the company on United States and foreign soil. This is further complicated by the necessity for CIOs to lead employees and outside contractors (e.g., offshore contractors) that have either deep technical expertise or deep business management competence (e.g., financial skills). Thus, he or she must be a leader of people and know how best to motivate and direct people to help ensure company success. The CIO should always hire people who are smarter than him or her. He or she should think first, plan second, and act last, avoiding shooting from the hip and regretting it later.

Challenges of a CIO

In lieu of communicating simply and easily, the next most challenging aspect is keeping up with the trends and changes in technology. To best address this challenge requires the CIO to have a strong sense for "where" technology is heading both within and outside their industry and to be well versed in the new technologies. The CIO has to know at a basic level what technologies are available, how they interact, how they can be integrated, and how they will influence the business and drive value. Ultimately, all IT value must be captured in financial terms. As a member of the executive team, the CIO must be able to speak in financial terms and articulate how technology applies and aligns with the enterprise strategy. In addition, when speaking with executives, investment analysts, or others outside the firm, he or she must be able to explain how technology investments will improve business profit, return on equity, return on assets, and in other terms used by Wall Street rather than simply explaining how technology works. A further challenge lies in leadership. More than ever before, leadership is critical for a CIO's success. According to Colin Powell, "A person is a true leader if people will follow him or her simply out of curiosity." For the

CIO, achieving this level of leadership is tough, considering the lack of technical standards, how quickly technology changes, and the degree of resource diversity in terms of competence and culture the CIO must face. Couple all this into leading large corporate initiatives with multimillion-dollar budgets that affect multiple business units sometimes across geographic boundaries, and the complexity of the challenge becomes evident.

Strategies of a CIO

There are two key areas a CIO must focus on to be successful: (1) the financials, which include the operating expense budget and the capital budget, and (2) the portfolio of technology projects that must be completed each year to enable profitable business growth. Each CIO must develop his or her own approaches or strategies that align with their specific business customers and company culture. However, there are some basic rules of thumb:

- From the business strategy and its required technology capabilities, construct an annual IT strategy and operating plan.
- Based on the results above, establish financial and capital project goals and objectives for the entire IT organization from top to bottom.
- Build a financial and capital project tracking and monitoring scorecard for each IT function (e.g., business intelligence department) and summarize for the entire department.
- Establish an online IT dashboard that tracks all projects IT is working on for the business (this assumes the business is prioritizing the projects). Make this dashboard available online so the business can easily access the information whenever necessary.
- Each month, conduct an IT operations review with all CIO reportees.
- Each week or every other week, conduct a program review with key project and program management people and conduct a detailed review of each project. Given the size of IT portfolios, a company may want to review a different section of the project portfolio at each session.

- Most importantly, be candid, ask questions, and develop action plans with a responsible party and due required dates, then hold people accountable.

Advice from a CIO

The best piece of advice a CIO can offer IT team members is to have patience and think through issues and problems with a questioning mindset before developing and delivering a solution. When either a new application is rolled out or a piece of technology changes and a team member must deal with a customer who does not understand the new change, the IT associate must remain patient and think critically about the problem's root cause and potential solution in the face of the customer's confusion and frustration. Most often, the fact is that the technology itself is the genesis of a customer's anger, not the IT associate. In addition, team members must work hard to communicate a technical topic as simply and clearly as possible.

Making Technology Decisions

A CIO should never make technology decisions alone. He or she should work closely with the business executive sponsor, the sponsor's team, and the CIO's own subject matter experts. In many companies, the CIO works side by side with new product development, operations, finance, and sales and marketing executives to identify, evaluate, and select new or improved technologies to help various products and services get to market. Yet, being effective with any of the executive management team requires the CIO to understand their pain points. Generally, this means understanding their goals and objectives, barriers to achieving those goals and objectives, and how applying technology can help. The CIO needs to build strong relationships with each executive knowing what the executive finds important and not important, and what they are passionate about regarding the business and how they see it succeeding.

Educating the Executive Team

Often, other executives do not necessarily understand the position of CIO or technology in general—some actually fear it. However, IT cannot be ignored, as it is now fundamental to business profit and success. All the executive team members must have a basic level of understanding of technology. Their knowledge does not have to be deep, but they must try to learn and educate themselves. In building his or her relationship with the executive team, the CIO can conduct brief education sessions to increase knowledge and understanding quickly. Once all members of the team are comfortable with technology, IT and associated business initiatives become easier to communicate, and the knowledge also helps to ensure better decisions and improves time to market. Another way to educate the executive management team is to buy each member a dictionary of technology terms and refer to the definitions whenever appropriate.

Setting Goals

When setting goals for the IT team, the CIO must first look at the annual business goals, strategies, and supporting operating initiatives. He or she should then work with the team (and, when appropriate, the business) to develop corresponding technology goals, strategies, and key initiatives. Once complete, the CIO and senior members of the team must present their annual business plan to the executive team, detailing operating cost, capital investment, and resource needs. Thereafter, periodically, usually monthly and quarterly, the CIO must provide an update on progress to the executive team. Throughout the year, each key IT initiative is reviewed weekly by the IT governance committee and must pass through a series of tollgates that measure progress and ensure that each project meets expected milestones, regulatory needs, costs, key deliverables, and timelines. Each month, key business executive sponsors and other managers are invited to participate in these meetings. This helps align the executives and IT since all project priorities are set by the business and reviewed accordingly.

Transforming a Company

Unless the CIO becomes the CEO, he or she must never be placed in a position to fully transform an organization. However, the CIO as part of an executive transformation team makes sense. The first step for any IT transformation begins with the business vision and strategic plan looking out three to five years. This foundation provides the CIO a means to create an IT vision and strategic plan that aligns with where the company must be in the near term and longer term. In all cases, the CIO must have complete support from the CEO and executive team, since he or she will have to restructure the IT organization, hire and fire employees, and change IT processes, applications, and infrastructure technology. To accomplish this appropriately requires the CIO to assess the current IT performance across the areas of people, process, and technology as well as how IT is interacting with its business partners. It is incredibly difficult to get everyone on the change bandwagon, reignite their passion, actively make changes, and understand how their contribution has improved the organization. Yet, as the team achieves at first small wins and then larger ones, it is very rewarding to see a team transformed and celebrating their successes.

Useful Resources

A CIO can use several resources to make technology decisions. The most important resource for any CIO is their various experiences. A CIO must keep abreast of technology through reading, research, conferences, meetings with peers, and networking. Valuable resources that tend to get overlooked are employees themselves. Not only should a CIO hire people smarter than him or herself, but the CIO should also listen to what they have to say.

Misconceptions about CIOs

The biggest misconception about being a CIO is many people believe a CIO should be engaged only on IT matters rather than leveraging their knowledge and experience for identifying opportunity outside the IT realm. A good CIO must understand how the firm makes money and where operational improvements in business processes exist with or without the use of technology. The CIO now tends to be a business executive that runs a business instead of just a technical person who advises the business.

Evaluating Technology

When evaluating technologies, a company must first determine the purpose, objective, and benefit it expects to receive from each. Some of the biggest technology expenses of an IT department are in the rebuilding and/or upgrading of current systems. Generally, the balance is targeted towards those that improve operations and customer service. However, the goal ought to be providing technologies that give the business new capabilities to compete, yet at the same time ensure that systems meet security, speed, and regulatory needs. Thus, a CIO is tasked with getting the right mix of technology while balancing cost and benefit. Once a technology is implemented, the IT team must run a post-implementation audit to determine whether the organization is getting the original benefits it intended from the technology.

CIO Golden Rules

There are three golden rules for being a CIO. The first rule is that technology must support and be aligned with the goals, objectives, and strategy of the business. The second rule is for both the CIO and the IT team to understand the nuts and bolts of new technology as well as any implications of future use it may have for the business. The third rule is that the CIO must leverage all sources of knowledge and experience—whether employees, vendors, or other sources—to find ways to reduce cost without adversely effecting internal or external customer service.

Richard Toole is senior vice president and chief information officer for PharMerica (a subsidiary of AmerisourceBergen). Mr. Toole serves a critical role assisting PharMerica in meeting its aggressive short- and long-term growth goals, especially in the area of strategic technology development and implementation.

Mr. Toole began his career with the Kimberly-Clark Corporation and then worked in information technology and strategy consulting with Ernst & Young and Accenture. During the late 1990s and prior to joining PharMerica, he was a vice president of strategic solutions with Time Warner Inc. and a consulting vice president with Gartner Inc.

Mr. Toole holds a bachelor's of arts degree in business administration from Michigan State University.

Dedication: *Dedicated to the PharMerica IT team.*

Technology Decisions: The View from a College Campus

Stephen H. Hess

Associate Vice President, Information Technology

University of Utah

Being a Technology Leader

Being a technology leader is about matching technologies and the changes brought about by technology with the business needs. A chief information officer (CIO) needs to be aware of new technologies, which keep coming at a faster pace. A CIO also needs to be very conversant with the needs of the clients he or she serves. Clients may be internal to the company, such as faculty at a university, or external customers in the case of an educational institution, the customer being the student. Many say needs should drive business change, not technology. I've found technology drives business change every bit as much as needs. Technologies like the Internet have changed markets, industry structures, services and products and their flow, consumer markets, consumer values, consumer behavior, jobs, and the internal workflow business process. These changes have added to the CIO role of "technology steward" the position of "business leader." The CIO is an information technology (IT) manager who provides vital IT support, but increasingly as a business leader provides insight into business vision and helps the company overcome barriers to change. In providing this direction, the CIO helps transform the business to better meet client needs.

A CIO should have three abilities. First, they need to understand the business they service. Only then can they provide IT services that will be helpful in creating new business opportunities. Also, company leaders and staff trust CIOs who have a good understanding of the business. Second, a CIO needs to understand technology and the changes technologies are making in the business and be able to explain them in non-technical terms. They can then give strategic advice on how IT will change the business in a way everyone can understand. Third, they need to be a leader who can match employee skills with projects and assist people in working together as teams toward timely completion of those projects. The CIO needs to be a team player themselves, a part of the business governance process, to suggest solutions, gain consensus, and help people make the necessary transition. They need to deal with political issues and clear away barriers to the successful completion of IT projects and meaningful business change. At a university, the CIO must understand higher education, preferably someone who is a part of the faculty, is respected on campus, and works with the deans and department chairs to bring about meaningful change.

To ensure that IT projects meet needs, keep within budget, are completed in a timely way, bring about new business opportunities, have a return on investment (ROI), and are accepted by employees and clients, we follow a six-step process:

The Six-Step Process for Adopting an IT Service

1. *Needs Assessment*

The first step is needs assessment, to assess the end users' needs so the IT service provided will be useful to the client.

2. *Product/Service Definition and Management*

The IT product or service must be defined in scope and in the specific function or services it will provide. Service level agreements are written by the client and IT staff. They are written in language the client can understand. These agreements include metrics that measure the success of the IT service provided in meeting the needs specified in the needs assessment. The results will be in a written agreement to the client. These agreements help manage expectations, scope creep, and escalating costs.

3. *Business Case*

The business case is a financial summary of the service level agreement and services to be provided. It assesses the total cost of operation and ROI for both the business and the client. It outlines the change in business process and workflow that will occur when the new IT process is put in place and financial savings that will result. The business case outlines how much of the business process and workflow will remain the same and what parts will change. There are generally greater efficiencies from an IT perspective when a business adopts the business process of the IT vendor or service. But the business may be less willing to change its processes and may want the IT service to align with current business processes. A CIO needs to help the business understand this issue and write in the business case the process both parties are willing to accept. It is generally a compromise. The ROI will be greatly affected based on the decisions made matching IT services and business processes. If there is not a fundamental business process change, many times IT service and the related expense are layered on top of existing operation cost. The result can be additional expense rather than savings to the company. Business process change is often difficult. IT processes may eliminate or significantly change jobs as business processes are automated. It may cause them to be reassigned to other parts of the business. But if there is not good alignment between business processes and IT services, substantial efficiencies will not be realized. The business case provides a written and financial resolution with a projected ROI before the IT project or services are developed.

3. Project Management

If, after an analysis of the business case, management decides to proceed with the IT project, a project management leader is chosen to lead the team to complete the project. It's generally a good idea that the project leader be freed from operations to have the time to lead the project team. Many times, operation managers who are already 120 percent occupied in their current job are asked to lead a project team. They are so busy with day-to-day operation that they often can't find the time to complete projects on time. Our project managers are certified, and they have experience in project management, an understanding of IT operations, and can lay out a project plan in writing so the project is communicated to the team, who can then be accountable for their part of the project.

According to an extensive survey of over 13,000 IT projects, only 34 percent of them are completed on time and within budget. The average IT project exceeds its allocated financial resources by 43 percent and its original time schedule by 82 percent. This is generally the result of poor or no business cases and/or project plans.

5. Marketing/Training

Marketing and training alerts the people who will be affected by the new IT service and resulting change in business practice. People are then trained in the use of the new service and business processes. People need be aware of the new service, the timeline, and ways it will affect their job in order for the new implementation to be successful. Hopefully, they would have been involved in the needs assessment and the definition of the service level agreements. Communication brings motivation and greater acceptance for change.

Adequate training is needed so all parts of the new process are used efficiently by the employees. Training has a direct impact on ROI. The scope of training will involve both IT people involved with the IT service and employees involved in the new business process. Clients many times need to be informed and trained in the new IT services.

6. Evaluation

Once the above steps have been completed, an assessment of the project needs to be done. Success or failure of the project will be initially focused on the metrics in the service level agreements, client satisfaction, timely completions of the project, staying within budget, and finally a positive ROI.

If followed, these six steps bring about the alignment of IT services with business and client needs. They are not meant to bog down the process of change but to systematically bring about change, taking into consideration all the important variables. These steps are summarized in written form before a project is started. Depending upon the complexity of the project or service, the written summary generally doesn't need to be longer than one to three typewritten pages.

The six-step plan is a standard within the industry. It helps IT and businesspeople work together to match technology to business and client needs. Many times, an IT person may discover a cutting-edge technology and want to implement it based more on the excitement generated by the technology rather than the business opportunity, filling needs, and bringing a ROI. The six steps help a CIO get both IT and businesspeople involved. This process can save thousands and millions of dollars in short- and long-term IT, business, and client costs.

Getting the Best ROI

Assuring that you will have a good ROI comes from a well thought out business case. For example, if our university is going to change to tuition payment online, we would scope out what it would cost to put an online service in place. The business and workflow process would change, because tuition in the past has always been gathered primarily by going to the student services building and paying tuition to a cashier at a window. Students will now pay online and not in line. The cashiers are no longer needed in the numbers they once were, so they would be reassigned. If they were not assigned, the IT process and expense would be overlaid onto the existing tuition operation expense and be an added cost with less or no ROI. For the tuition collection process to be successful, there must be a change in the business process, which generally involves changing or eliminating people's jobs. There is also an ROI for the client, the students, who no longer need to come to campus with the cost of gas, time off work, parking, travel time, and time standing in line. The client ROI could be diminished if one or two steps in the process still required them to come to campus.

Adopting new technologies that are not bleeding edge helps to achieve a better ROI. We are rapid followers. We want technologies that have been proven in the marketplace. It is expensive to be a guinea pig. If other companies or universities want to be the test case, we will be happy to learn from their experience. We don't want to be the beta test of some vaporware. We want product that has been tested in a business environment much like our own. We call for positive references from people who have used the product.

That new technology must be in alignment with and fill our needs. It must meet performance requirements for the size of the project. The administration needs to support the project and the business processes that will change. The project must add business value. All these factors will improve the chances for a greater ROI.

For a business case to accurately predict ROI, our team needs to analyze and account for every expenditure, including the savings from a realistic shift in workflow and business process. We are just installing next-generation voice phones, an investment of about $10 million. We want to make sure we account for every expense and how they will save people time and money. We make sure people are aware and trained on the systems so they will be fully utilized. If they're not trained and aware, they may not even know these features exist in the system, and they won't take advantage of them.

The best advice I've received is to use ROI tools that capture constituent benefits along with internal returns. Education exists to serve students and faculty, so those are the benefits that really matter. Benefits are what ultimately drive our agency budgets. Then we need to identify and keep within scope the initial needs identified. That means we don't let the project become bloated and cover too many things. Cover what you initially send out to take care of needs, and stick with that. Find systems that are inexpensive to maintain and connect with your other systems. Consolidate the equipment, streamline the business process where you can, and redeploy staff. Again, you have to have the support of the central administration to do that. We give priority to the "low-hanging fruit" projects that deliver the greatest benefit sooner than later at the lowest cost.

Resources and Expenses

Facts are my most useful resource. You need to know if something is a reliable technology, if it's proven in the marketplace, if it can speak to different networks, and if it's open-sourced and not too proprietary. You also need to understand and know the total cost of ownership. You need these facts to make good technology decisions.

Our biggest technology expenses are networks, telephones, campus-wide IT applications, Internet connectivity, and security.

With any IT project, we've developed spreadsheets to calculate ROI. We identify the cost of implementation, capital, operation, and replacement. We factor in the business process changes that will be made to bring about greater productivity and efficiency in dollar amounts. All of these costs are brought together in the business case to make a better technology decision. It is tedious and takes some time, but like everything else, the more you do it the more skilled you become and the task becomes easier. We have a business case person who looks at all the ROIs on all the technology services we provide. Part of figuring an accurate ROI is making sure one service doesn't subsidize another. For that reason, we un-bundle costs. We then have a clear ongoing assessment that an IT service is making the projected ROI. We run our IT organization as an internal business. If we had to compete, would it be cheaper to outsource than for us to run that particular application? In the process of determining whether an IT service is making the projected ROI, we have outsourcers bid on our IT services to see how we compare. Again, all of this information is put into a dollar spreadsheet and analyzed to reach the best decision.

The Goals of the Team

We set goals for the team through strategic planning. That process is a thorough one. It involves an environmental scan. We try to predict the future environment in a number of areas: our clientele, technology, the economy, revenues, competition, and so on. Part of the scan includes looking at SWOT (strengths, weaknesses, opportunities, and threats). After an analysis of the environmental scan coupled with a needs assessment of our clients, we set our mission, vision, values, and goals that will steer us in a direction so the company will be strategically placed in the future environment to be relevant and fit the needs of our clients. Companies that don't pay attention to these future forecasts may miss important course corrections and lose clients. In a rapidly changing environment of IT, a company can go into what is termed freefall, losing market share to more savvy competitors. Within goals we set, there are several specific IT projects we want to complete for the year and operation improvements.

There was a great article in the *Harvard Business Review* about decisions IT people should not make. Some of the IT decisions executives should make include: how much should the company spend on IT, which business processes should receive our IT dollars, and which IT capabilities should be firm wide. We take the six-step process for adopting an IT service and explain to management what we plan to do. The six-step process and our plans give them what they need to know in order to make decisions on how much money to spend, whether this really fits within the strategic goals of the university, and whether it will help the university in completing its mission. Generally, that's the kind of discussion we have.

At a university, a further challenge is making sure you touch base with all affected using parties and that there is a consensus that this will be the tool that is used. The second piece would be making sure that when you purchase these systems, there is a good purchasing process in place. We go out to bid. If you go out to bid, you will get lower prices. There's no such thing as a good partnership with a vendor when it comes to pricing. They're out to make money, and you need to make sure they sharpen their pencils.

Stephen H. Hess, Ph.D., has worked in educational and information technology services for more than thirty years. He has also worked in business, taught in the public schools, and teaches at the University of Utah.

At the University of Utah, Mr. Hess has been director of instructional media services, director of the university press, general manager of K-ULC TV, assistant vice president for student and university relations, executive director of media services, the first executive director of the Utah Education Network, and assistant academic vice president for electronic communications. He also holds adjunct faculty appointments in the Graduate School of Education and the communication department.

In his current administrative assignment, Mr. Hess oversees the strategic direction for campus information technology. He has served on several national, state, and campus committees dealing with education and information technology and also chaired the state legislative planning task force on education technology, which developed one of six action plans for higher education.

Mr. Hess holds a doctorate in educational administration with an emphasis in higher education, a master's of education in instructional systems and learning resources, and a bachelor's degree in history with a minor in psychology. He is widely published and presents nationally. In all his work, his primary motivation has been to help improve education with the use of technology for the benefit of students and people.

Corporate IT Governance

James F. McDonnell
President
McDonnell Consulting Group LLC

Integrating Technology and Business Needs

In order to be successful, a chief information officer (CIO) must be able to integrate technology with business needs and provide the leadership required to ensure that the information technology (IT) department understands the strategic direction of the company and that the other executives understand the application of technology to enhance business processes. Our company uses a very formal IT governance process that integrates technical and business ideas. Each business line manager assigns a representative to the IT advisory committee who is responsible for representing the technology needs of the particular business line and helps assess the needs for enterprise architecture enhancements. Each member of the committee submits business line-related problems and initiatives they think IT or technology might assist with. With all of the business line needs identified, the committee then attempts to prioritize those needs and determine the return on investment. The IT director then takes the committee report and assigns members of the IT department to look for technology solutions and report back with costs and schedules. The IT advisory committee then reevaluates the costs, schedules, and returns on investments to develop a report with recommendations for technology investments and submits their recommendations to me. I review the findings with the other executives within the company, and from there we decide whether we want to invest in a particular technology. We do this on an annual cycle.

The technology solutions we currently deploy are enterprise architecture solutions (like e-mail and normal desktop applications and hardware) and custom applications that support inventory control, sales and marketing, business scenario processing tools, and digital control systems for two nuclear plants. We are moving to commercial, off-the-shelf technologies whenever possible to enhance data integration and reduce development costs.

My Value Added

The most obvious way I add value to the company is by maintaining a secure and reliable IT backbone. As an officer of the corporation, I also participate in corporate strategic planning, which allows me to ensure that the IT investments are strategically aligned with our business direction. My final major added value is to ensure regulatory compliance with the Securities and Exchange Commission, the Nuclear Regulatory Commission, and the Department of Energy.

The CIO must be a business leader first. IT provides a means to an end, but cannot be the end in itself. Understanding the business and business goals must be the primary focus of the CIO. He or she needs to know how to manage the acquisition and application of technology, but must always keep in mind that the technology must be needed to support a core business requirement.

A CIO must have an understanding of the business planning cycle, budgeting and financial planning constraints, and long-term investment strategies. The CIO must also have an understanding of return on investment and the technical capability of IT department leadership and staff. I use the IT governance process as described above to communicate the process throughout the company, and I use broadcast e-mails, town hall-like meetings, and staff meetings with our peers in management. I also conduct a formal customer survey each month with each executive to review specific IT issues related to their operations. The feedback from these meetings is fed back to the IT advisory committee for consideration.

CIO Challenges

The IT world is constantly changing, and IT organizations have to change with it. Some talent that was required five years ago is no longer required. Professional development and strategic planning is critical. We are using fewer custom applications and more commercial, off-the-shelf applications and will most likely continue this trend. The IT department in a non-IT company is becoming more focused on information integration and information integrity. This is a big change, and it makes people nervous; the CIO's leadership challenge is to keep good people doing challenging work in often new and different directions. Leading people and leading change, and being comfortable with change, is as important as understanding the business operations. The IT world is constantly moving, and we must keep people aware of that, and they need to be comfortable knowing they will be able to change with it. I tell my team that if you are coming to work two days in a row and doing the same thing, that's okay, but if you are coming to work two years in a row and doing the same thing, something is wrong.

Working with Other Executives

I work closely with all of the other executives. That being said, I probably work most closely with the chief financial officer (CFO) right now because of the importance of internal controls on financial systems. We speak routinely about the issues related to maintaining financial data security and ensuring that all of the data that feeds into financial systems is protected by controls that ensure data integrity while simultaneously keeping large amounts of information flowing. I need to know when there are problems with the balance between data protection and data use, and I need to make sure that when changes are required the IT department understands the need and relative priority. This interaction is critical to making sure we are focusing on business needs and not just the desires of some individuals.

Changes in internal controls at the enterprise level invariably affect individual activities at the desktop, and as a result the CFO-CIO relationship needs to be able to keep the direction at a strategic level. The financial operations of the business are highly dependent on IT systems and staff. The enactment of Sarbanes-Oxley and the resulting changes in the audit processes have put significant pressure on the CFO's office, and the integrity of data is a critical component of a complex system for which the CFO is accountable. When I have discussions with the CFO, they are normally followed up with a discussion with my corporate IT director. Since I expect the IT staff to always be looking for new and creative ways to solve problems and make us more efficient, it is imperative that the IT leadership has a detailed understanding of the issues confronting the firm. The IT staff input is technology solutions, and they need to be looking to the future all the time. Meanwhile, the business minds need to be articulating what the business needs are, so IT can look for solutions that fit those needs.

Upon my arrival about a year ago I established an IT governance structure for the company. I worked with our internal and external auditors and other executives to create a structure that ensures that IT investments are in fact in line with business needs. That is probably the most important IT activity that has happened in our company, and it changes the nature of the IT organizations. IT is no longer a technology group providing tools for people. Instead, IT is about ensuring that the data needed to manage the company gets where it needs to be in the right format at the right time and with a very high degree of reliability. Another step that was taken right after I arrived was to integrate the IT departments across the business lines. When I first got here, we had four IT departments, each reporting to a business line manager and each developing their own plans and budgets. The obvious result was a lot of duplication of effort, redundant systems, and a talent pool that was not properly aligned with business needs. I promoted one of the IT managers to a newly created corporate IT director position and put the business line IT managers under a signal director. We then segmented the enterprise systems from the business line systems to ensure that services that are cross-cutting are centrally managed while preserving the ability of a particular business unit to have custom applications as appropriate. This approach has given us a lot more efficiency in managing the enterprise and has streamlined decision making and project implementation.

Working with My Team

When we set goals for my team, we begin by referring back to defined business technology requirements. There is a business requirements timeline for implementation of a new technology that ultimately drives the technology development and implementation schedule. If a plant general manager requests a new application to be designed and installed, he articulates the need through his IT advisory counsel member and the governance process I have already described establishes the macro goals. Once that project is approved, a customer timeline and a budget are developed. I then review the budget, timeline, IT infrastructure requirements, and financial and resource constraints with both the corporate IT director and the enterprise IT project office manager. We agree to those milestones up front and a detailed project plan, including goals, is implemented. I use a series of weekly dashboard-like reports for each of the departments; one of those reports is the IT project implementation status. I can see a snapshot each week of the status of all the major projects underway, including costs and schedules. With this tool, I get a list of every project underway and where they are in relation to the schedule, and I can make adjustments to goals and milestones if required. Another important part of this process is working with other business leaders to make sure I understand the possible unintended consequences of changing a business process. The corporate IT world is one integrated system of systems; nothing can be done that doesn't have some impact on other parts of the system. A key goal in every project is identifying those interdependencies and ensuring that we have milestones that require looking for those unintended consequences whenever changes are being implemented.

As a general rule, I believe IT changes are often made too fast. There is a massive marketing effort underway by IT companies, and there can be a tendency to see a new technology and look for a place to use it. This can result in technology driving needs, not business driving needs, ending up with a system that is not properly integrated and planned out. I often tell my team members: Don't buy technology because it's new and cool; buy it because it will save us time and money or better protect the integrity of our information.

Our spending decisions are made through the IT governance process. The largest expenditures are maintaining the enterprise systems: the large financial systems, inventory control systems, and so on. The level of effort that goes into the existing systems is defined in a service level agreement signed by the chief operating officer and myself. This agreement defines goals and metrics for maintenance of the enterprise systems, which accounts for most of the IT department work. Only about 20 percent of our IT budget goes toward totally new initiatives; most is spent maintaining, operating, and updating the current systems.

Technology Decision Making

Conducting due diligence on a technology purchase begins with an identified need that has to be validated by a business line leader; this doesn't come from the IT department. The IT department then matches technology and runs the technology in a test environment. We will rarely buy something for system-wide implementation. Instead, we buy a test version and run it on a test system. For example, when we were looking into an upgrade of our e-mail system, we tested at one plant for about six months before we decided to expand throughout the company. In the case of critical systems that affect data integrity, we will test offline for a period of time before moving to our production systems. After the test period, we review the potential risks of moving into a production system or going enterprise-wide. We ask: Does it change any business processes? Does it affect any critical systems? Are there other applications that are linked to it that we haven't looked at? There is a very elaborate vulnerability assessment that has to be done before we decide to launch a new technology.

In our case, some of the most difficult decisions are the ones that affect business-line operations. We run two nuclear facilities that operate in what are known as safety-conscious work environments; we have very elaborate processes and procedures in place, built on decades of safe operations. When anything we do changes one of those procedures, it has to go through a very elaborate change review process that addresses safety impact. This is separate from and in addition to all of the IT department reviews. The change control process is very important and a high priority for us, but the approval process can make it very difficult to do a timely change in our IT enterprise. Understanding this limitation is part of the strategy for segmenting those business lines from the enterprise architecture itself. We work hard to make sure the system integration does not result in enterprise systems and does not affect nuclear safety-related processes. That is a core responsibility for our department: to really make sure we absolutely understand if there are any safety implications in addition to understanding the impact on other IT systems.

I think the biggest misconception about our department is that our decisions are about technology alone; instead, they need to be about business. The IT management structure is still relatively new in corporate America. IT has changed so much in just a short amount of time that there is still a perception that IT is about technology and not about the business; it needs to be the other way around.

A Look at Return on Investment

The primary measure of return on investment I like to use is increased efficiency and the associated reduction of labor costs. Technology should save people time, so we measure how much time we save. If one is not careful, we risk adding more requirements on people to develop, install, and maintain a new system without a payback. One of concerns I always have in looking at new proposals is whether we are actually adding work without an incremental offset.

We have an annual review of the effectiveness of our technologies. We ask:

- Is the technology doing what we said it would do?
- Is the investment requirement estimate right?
- Did we in fact become more efficient?

The process of identifying a need, identifying the technology to meet that need, and installing the technology is only completed when we follow through and measure results. We must examine whether the change accomplished what was expected and how well we did it. A technology is only a worthwhile investment for our company if it results in cost savings or increased data integrity.

Top Resources

The most useful resources in my position are other people. I talk to a lot of people: staff experts, customers, and peers in industry. I also read a lot. There are a lot of good ideas and experiences in print, and the IT world is too complicated for any of us to personally experience everything. I look for other people who have similar problems or issues and see how they implemented things and how it went. If I talk to a vendor, I will ask them to show me successes and failures; I always want to know why something didn't implement correctly.

Talking to people helps ensure that this job stays a business leadership job, not an IT job. It ensures that I continue to know the business line our technology supports. Technology leaders must understand a demonstrated need, understand return on investment factors, and be able to say no, and that all begins with knowing your peers and knowing your business.

James F. McDonnell is president of McDonnell Consulting Group LLC, a Virginia-based firm that specializes in strategic planning and assistance for small companies in the homeland security business and providing risk management advice to large firms. Until September of 2005, Mr. McDonnell was vice president and chief information and security officer for USEC Inc., a Fortune *1000 global energy company. He was responsible for the strategic direction of information technology operations, physical security operations at several nuclear facilities, and information security programs at the company's headquarters and subsidiaries.*

Mr. McDonnell has nearly thirty years of management and leadership experience in various segments of government and the private sector. Immediately prior to joining USEC, he was director of the protective security division in the Department of Homeland Security. As part of the original executive leadership team there, he was responsible for the initial development and management of what later became the infrastructure protection office. Mr. McDonnell conceived and directed the development of several national programs at the Department of Homeland Security, which included management of security activities through several high-threat periods, the development of a national protection plan, and implementation of the buffer zone protection plan. The protective security division received the "Excellence Award" at the secretary's First Annual Awards Ceremony for many of the accomplishments that took place under Mr. McDonnell's leadership.

Mr. McDonnell received his bachelor's of science degree from Regents College, University of the State of New York, and earned a master's of arts degree from Georgetown University. He also attended Harvard University's senior executives in national and international security program. He is a member of the United States Chamber of Commerce homeland security policy task force, a senior fellow at George Washington University Homeland Security Policy Institute, a member of the CSO Roundtable, and sits on the board of advisors for Digital Sandbox Corp.

C-LEVEL BUSINESS INTELLIGENCE
ASPATORE
BOOKS
C-LEVEL BUSINESS INTELLIGENCE